551.697
HAR

CO-AZM-593

37512000148305

18.95

6-10-85

DATE DUE

AUG 12 '85			
SEP 30 '85			
APR 01 1986			
SEP 2 8 1988			
AP 1 7 '93			

DEMCO 38-297

WITHDRAWN

BLUE NORTHERS TO SEA BREEZES

Texas Weather and Climate

NESBITT MEMORIAL LIBRARY
529 Washington
Columbus, Texas 78934

Hendrick-Long Publishing Co.
Dallas, Texas

Hendrick-Long Publishing Company gratefully acknowledges the Forest Service, U.S. Department of Agriculture,
for permission to use illustrations from the book, FIRE WEATHER.

Design: Ellen McPeek

Copyright © 1983
Hendrick-Long Publishing Co.
P.O. Box 12311 • Dallas, Texas 75225

ISBN 0-937460-10-9

CONTENTS

Preface

Meteorology is a science; weather forecasting is a profession. Whereas the ability to forecast the behavior of a physical or chemical system is the goal of most scientific inquiry, it is particularly relevant in meteorology because of the general interest in day-to-day weather. However, it is obvious that the ability to forecast weather accurately and consistently has not been realized.

The secret to better forecasts is a better understanding of atmospheric behavior. This book presents an up-to-date synopsis of meteorology and the physical laws upon which it is based. It is one thing to describe a tornado and quite another to understand its origin and development.

The first three chapters discuss the nature of the atmosphere independent of geographic location. In chapters 4 and 5 the focus shifts to Texas and a description of its weather and climate based upon the causes and effects presented in previous chapters. It is hoped that this format will encourage the reader to pursue the science in an attempt to explain the weather events and data presented. Meteorology is particularly well suited to beginning studies in science since it relates to a student's first-hand experiences through constant contact and interaction with the weather.

As our understanding of atmospheric behavior improves, so will weather forecasting. The profession (weather forecasting) is dependent upon the science (meteorology).

Chapter 1 / THE AIR AROUND US

It's been said about Texas weather that if you don't like it at a particular time, wait around a few hours and it will change. Certainly there is truth to this statement, but the implication that the changes are random and unpredictable is invalid. The atmosphere behaves according to fundamental physical relationships and laws. The inaccuracies in weather forecasts result from our incomplete understanding of these relationships.

In order to understand Texas weather, it is necessary to consider its underlying causes. Once this is accomplished, weather events which occur anywhere on earth (or on other planets as well) can be viewed in terms of cause and effect. This knowledge can then be used to forecast the weather. The complexity of atmospheric events is so great, however, that a weather forecast can never be more than a statement of probabilities.

The purpose of this book is twofold: first, to present an overview of the science of meteorology, and second, to apply this information to Texas weather and climate.

INTRODUCTION

The origin of the earth's atmosphere, like the origin of the solar system itself, is still the subject of much speculation. One thing appears certain, however. Because the composition of the air has changed significantly with time, the atmosphere we observe today is not the same as that when the earth was formed or during subsequent stages in its development.

Assuming the sun and the planets in the solar system have a common origin, it is likely that the planets' original atmospheres were alike. As the planets cooled, gases were released from their interiors and were prevented from escaping by the force of gravity. The larger planets had stronger gravitational attraction which enabled them to retain their original atmospheres. Smaller planets with less gravitational pull were at a disadvantage and tended to lose some of the lighter gases in their atmospheres to interplanetary space, a process which is continuing even today.

Another important factor is planetary temperature.

Molecules in a gas move faster as the temperature is increased and consequently have a greater probability of escaping to space. Thus, the planets best equipped to retain their original atmospheres were large (strong gravitational attraction) and cold (farthest from the sun). Small planets near the sun were most likely to lose the lighter gases which rose to the top of the atmosphere and subsequently escaped. An example is Mercury, the small planet closest to the sun which has lost all but the heaviest gases from its atmosphere. Since the time of origin, the atmospheric composition of small, hot planets has changed most significantly, while large, cold planets have changed very little.

Figure 1.1 illustrates the location of the planets with respect to the sun and their relative sizes. Since the diagram cannot be drawn to scale, the distances between the planets are not indicative of the actual distances in space. These distances, along with other pertinent data, are shown in Table 1.1.

ORIGIN AND COMPOSITION OF THE ATMOSPHERE

1

FIGURE 1.1
Planets in the Solar System and
their Relative Sizes

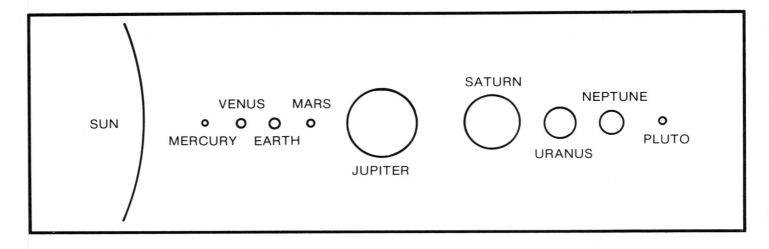

Distances in Table 1.1 are expressed in both miles and kilometers. The mass* of each planet is expressed in relation to the mass of the earth which is taken as one. Thus, Venus is 82% as massive as the earth while Jupiter is 317.9% or 3 times as massive. The sidereal period of a planet is the interval between two successive returns of the planet to the same point in space, as seen from the sun. It expresses the time required for the planet to complete one revolution of the sun. For the earth, the time period required is 365¼ days (1 year). A complete revolution for Pluto requires 248 years while Mercury makes a complete cycle in only 88 days.

The inner planets, Mercury, Venus, Earth and Mars, are known as the "terrestrial planets," while the outer planets, Jupiter, Saturn, Uranus, Neptune and Pluto, are called the "Jovian planets." It is obvious from Figure 1.1 that, with the exception of Pluto, the Jovian planets are substantially larger than the terrestrial planets. Table 1.1 indicates that they are considerably more massive as well. Since they are farther from the sun, the atmospheres of

the Jovian planets should also be cooler and thus better able to retain their original atmospheres. In fact, the atmospheres of extremely massive planets such as Jupiter and Saturn are probably much the same today as they were near the time of origin. On the other hand, the terrestrial planets are warmer and less massive suggesting a significant depletion of the lighter gases during the life of the solar system.

The early atmospheres of the planets were probably composed chiefly of hydrogen (H), the lightest and most abundant gas in the universe, helium (He), and compounds of hydrogen such as methane (CH_4) and ammonia (NH_3). These gases have been lost in large part by the terrestrial planets but retained by the more massive Jovian planets. Observations indicate that both methane and ammonia are presently important elements in the atmospheres of Jupiter and Saturn. Therefore, each planet developed an atmosphere which is unique in the solar system, depending on its mass and temperature.

The earth's present-day atmosphere probably did not

---

*Mass is a measure of the quantity of matter and does not depend upon the location, size or shape of a particular object. The more mass something has, the more matter it contains.

TABLE 1.1
Basic Data on the Planets

Planet	Mean Distance From Sun		Mean Diameter		Mass with Respect to Earth	Sidereal Period of Revolution
	Millions of km	Millions of miles	km	miles		
Mercury	58	36	4880	3031	0.055	88 days
Venus	108	67	12112	7523	0.82	225 days
Earth	150	93	12750	7919	1	365¼ days
Mars	228	142	6800	4224	0.107	687 days
Jupiter	778	483	143000	88820	317.9	12 years
Saturn	1427	886	121000	75155	95.2	29 years
Uranus	2874	1783	52000	32298	14.6	84 years
Neptune	4504	2794	48600	30186	17.2	165 years
Pluto	5900	3670	6000	3727	0.01	248 years

develop until much later when the original atmosphere was lost. As the earth cooled, gases such as carbon dioxide, nitrogen and water vapor which had been dissolved in the hot, melted rock were released. As cooling continued, the water vapor condensed to form the oceans. The water in the oceans then gradually absorbed most of the atmosphere's carbon dioxide, leaving nitrogen as the predominant gas in the atmosphere. Oxygen appeared as a result of "photosynthesis" after primitive plant life had developed. The process of photosynthesis, which is the formation of carbohydrates from water and carbon dioxide in plant tissues exposed to sunlight, releases oxygen as a by-product into the atmosphere. The present composition of earth's atmosphere is given in Table 1.2. Figure 1.2 illustrates the relative abundance of nitrogen and oxygen.

In addition to the natural constituents in the atmosphere, large quantities of potentially hazardous gases are introduced into the air as a result of human activity, primarily automobiles and industrial processes. Most prominent among these are carbon monoxide [CO], sulfur dioxide [SO_2], nitrogen oxides and hydrocarbons. The at-

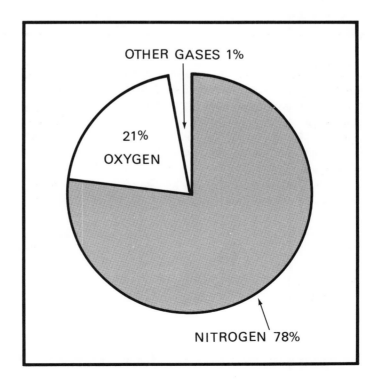

FIGURE 1.2
Atmospheric Composition

OTHER GASES 1%

21% OXYGEN

NITROGEN 78%

3

TABLE 1.2
Composition of the Atmosphere

Permanent Constituents		Variable Constituents	
Constituent	Percent by Volume	Constituent	Percent by Volume
Nitrogen	78.084	Water Vapor	4
Oxygen	20.946	Ozone	0.07×10^{-4}
Argon	0.934	Sulfur Dioxide	1×10^{-4}
Carbon Dioxide	0.032	Nitrogen Dioxide	0.02×10^{-4}
Neon	18.18×10^{-4}*	Ammonia	Trace
Helium	5.24×10^{-4}	Carbon Monoxide	0.2×10^{-4}
Krypton	1.14×10^{-4}	Dust	10^{-5}
Xenon	0.087×10^{-4}	Water (liquid & solid)	1
Hydrogen	0.5×10^{-4}		
Methane	1.5×10^{-4}		
Nitrous Oxide	0.5×10^{-4}		
Radon	6×10^{-18}		

mosphere also contains large numbers of solid and liquid particles, principally soil and dust, salts from the evaporation of sea-spray, smoke from combustion and ash from volcanoes.

In late March and early April, 1982, a volcano known as El Chichon erupted in a remote corner of southeastern Mexico injecting huge amounts of ash and sulfer dioxide into the atmosphere. A large percentage of this ash will remain in the atmosphere for a year or more and lesser amounts for two or three years. In addition to producing some spectacular sunsets, it is probable that the ash will cause a reduction in incoming solar radiation which will cool the lower atmosphere. Since many factors are important in shaping the weather during a given year, the eruption of El Chichon does not imply that cold winters and cool summers are inevitable; however, other factors not considered, the odds are tilted in favor of slightly cooler weather for awhile.

*10^{-4} indicates that the decimal point must be shifted four places to the left. Therefore, 18.18×10^{-4} is equal to .001818.

In order to understand atmospheric temperature structure, it is necessary to consider the distribution of energy. Heat energy can be transmitted from one place to another in three ways: conduction, convection, and radiation.

"Conduction" [Figure 1.3] is the means by which one end of a metal rod becomes hot when heat is applied to the opposite end. The heat energy is transmitted by point-to-point contact of neighboring molecules which remain fixed in their position. Although heat conduction in the atmosphere is too slow to be of consequence, it may become important when considering the exchange of heat between the earth's surface and the air in contact with it.

"Convection" occurs in liquids and gases which are free to move about. As the fluid moves, it transports the heat energy which is stored within it. It is common knowledge that warm air rises indicating an upward transport of heat by convection. Heating a kettle of water sets up convection currents which transfer the heat throughout the water as shown in Figure 1.4.

"Radiation" is the transfer of heat energy without the involvement of a physical substance to transmit the heat. Thus, heat may be transferred through a vacuum (space with no matter present) by radiation. It is the radiative process which transfers energy from the sun to the earth. How this energy heats the earth and its atmosphere is important to the understanding of temperature structure.

It is common knowledge that when a person is exposed to direct sunlight (radiation), skin temperature increases. This is due to the skin absorbing a portion of the radiant energy. The absorbed radiation is converted to heat and the skin temperature rises accordingly. The intensity of the radiation depends upon the distance from the source as shown in Figure 1.5 and upon the angle at which the radiation is received. Figure 1.6 illustrates how the radiation is spread over a larger surface when the sun is low on the horizon and strikes the earth at an angle. When the sun is overhead, the energy is more concentrated and more intense.

TEMPERATURE STRUCTURE OF EARTH'S ATMOSPHERE

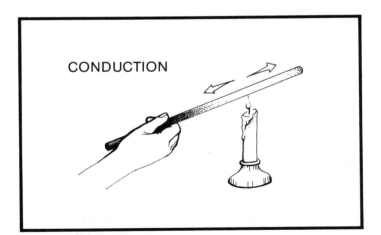

CONDUCTION

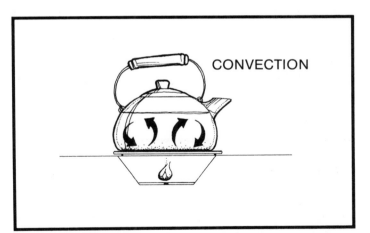

CONVECTION

FIGURE 1.3
Heat Transfer by Conduction

FIGURE 1.4
Heat Transfer by Convection

FIGURE 1.5
Heat Transfer by Radiation

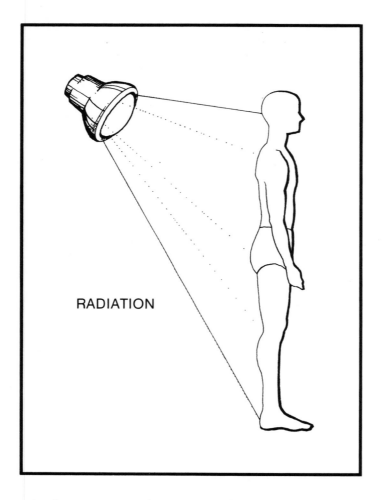

RADIATION

In the same way that a person's skin is heated, the earth and atmosphere are also heated by the absorption of solar radiation. The radiation shown in Figure 1.6 is absorbed by the earth's surface and serves to heat the earth. Since a large portion of the solar radiation passes through the atmosphere and is absorbed in this way, direct radiation is important in controlling the earth's temperature. The radiation that passes through the at-

mosphere without interacting with the air is said to have been "transmitted."

However, all of the radiation which is incident at the top of the atmosphere is not transmitted. A portion of the energy is absorbed by gases in the atmosphere. The most important effect is the strong absorption of "ultraviolet" radiation from the sun by the gas "ozone" in the upper atmosphere. This produces a high-temperature region in the atmosphere at an altitude of approximately 50 km where the absorption effect is maximum.

Ultraviolet radiation is that portion of incoming solar radiation that causes sunburn. If much of this radiation was not absorbed in the upper atmosphere and was allowed to penetrate to the earth's surface, the result could be harmful or even lethal to life on earth. This has created concern in recent years after it was suggested that certain chemicals used as propellants in hair sprays, deodorants and other common household products were interacting with the ozone in the atmosphere and causing a decrease in ozone concentration. This would allow more of the harmful radiation to penetrate to the earth's surface. Whether this is actually occurring to any significant extent is under investigation.

As a consequence of the absorption of radiation both at the surface of the earth and in the upper atmosphere, the vertical temperature structure is characterized by two temperature maxima, one at the surface and the other near 50 km altitude as shown in Figure 1.7.

On the basis of this vertical temperature structure, it is convenient to divide the atmosphere into four layers. The region below approximately 10 km in which temperature decreases with height is called the "troposphere." It is the troposphere that contains, with few exceptions, all of the clouds and weather which we experience. This is due to the fact that water vapor, the source of all weather, is confined to this lowest region. Above the troposphere is the "stratosphere," the layer in which temperature increases with height as a result of absorption of radiation in the atmosphere at approximately 50 km altitude. Above

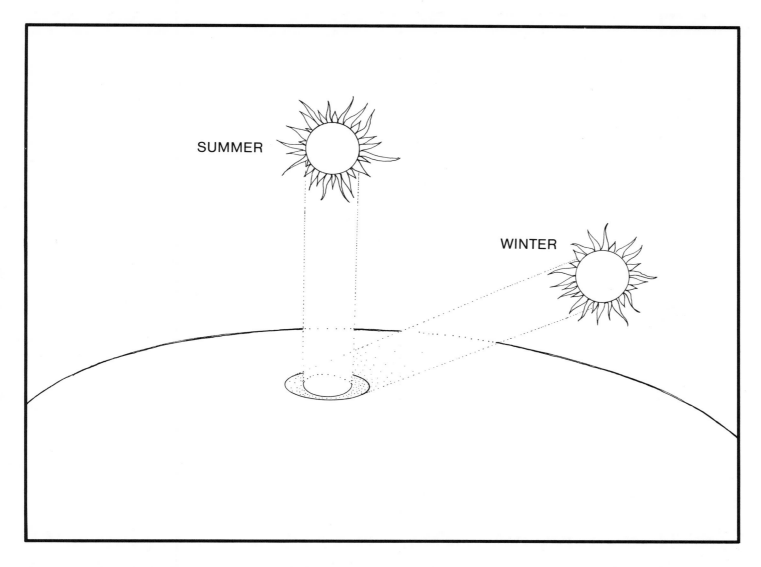

FIGURE 1.6
Incoming Solar Radiation
in Winter and Summer

the stratosphere is the "mesosphere," a region in which temperature once again decreases with altitude. The uppermost layer, the "thermosphere," is a region of increasing temperature but is of little consequence to weather events in the lower atmosphere. Boundaries between layers are the "tropopause" at approximately 10 km, the "stratopause" at 45 to 50 km, and the "mesopause" at approximately 85 km.

7

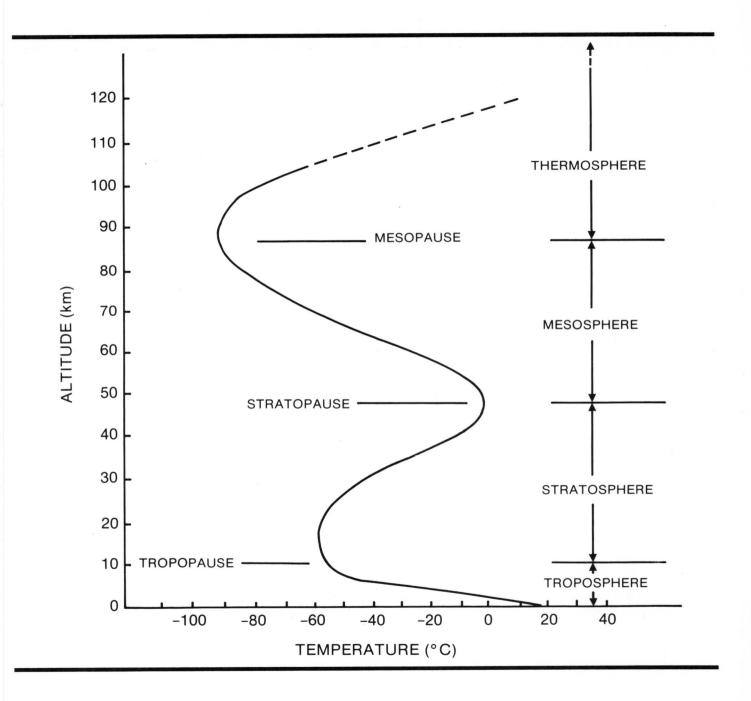

FIGURE 1.7
Vertical Structure of the
Atmosphere Based on the
Change of Temperature with
Height

Chapter 2 / OBSERVING THE WEATHER

In the same way that a person is described according to height, weight and complexion, the atmosphere at a given time and place may be described according to physical variables called "variables of state." However, measuring the state of the atmosphere is not easy. It requires accurately-calibrated instruments rugged enough to withstand weather events, and proper interpretation of data based upon an understanding of meteorological processes. The variables conventionally measured are temperature, pressure, wind speed and direction, humidity, clouds and precipitation.

Temperature

Probably the most familiar variable among the general public is temperature. We dress and heat and cool our homes according to the forecasted temperature. But temperature is much more than a measure of hot or cold. It is the property which determines the direction and magnitude of the transfer of heat energy in the earth-atmosphere system. If two adjacent bodies, such as the ground and the air in contact with the ground, are at the same temperature, there is no heat transferred between them. On the other hand, if one body is warmer than the other, heat is transferred from the warmer body to the cooler one. The flow of heat will continue until the two bodies have the same temperature, a condition known as "thermal equilibrium." Thus, temperature differences between regions in the atmosphere or between the atmosphere and the ground, govern the direction and quantity of heat flow. This process reflects nature's attempt to establish equilibrium by transferring heat from hotter areas to colder areas. The rate of heat transfer depends upon the magnitude of the temperature difference.

Heat transfer often takes the form of a "convection cell" [Figure 2.1]. As illustrated by the arrows, warm, light air is rising at the equator and cold, heavy air is sinking at the poles. The cell is completed by the transfer of air from pole to equator at low levels and from equator to pole at upper levels. The same principle produces a "sea breeze" which occurs along a coastline, primarily during the summer. Because land heats more rapidly than water, the air over the land becomes warmer than the air over the water. As a result, the warm air over the land rises and is replaced by cooler air flowing onshore from over the water. The cycle is completed by downward moving air over the water which replaces low-level air flowing onshore [Figure 2.2].

ATMOSPHERIC MEASUREMENTS

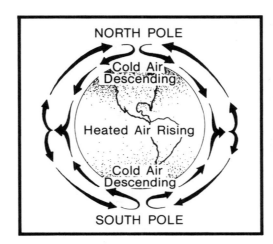

FIGURE 2.1
Simplified General Circulation
of the Atmosphere

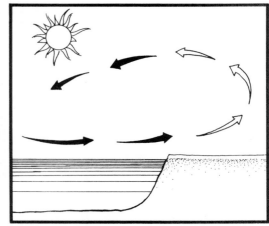

FIGURE 2.2
Land-Sea Breeze Circulation

Temperature is commonly expressed in three different systems of units, the most familiar being the Fahrenheit system. This scale defines the boiling point of water as 212°F and the melting point of ice as 32°F.* Temperature may also be expressed in degrees Celsius or centigrade. In this case, the boiling point of water and the melting point of ice are defined to be 100°C and 0°C, respectively. A comparison of these scales is shown in Figure 2.3.

The third scale, used primarily by scientists and engineers, is the Kelvin or absolute temperature scale.

This scale is based upon "absolute zero," the point at which temperature is so low that all molecular motion ceases. This occurs at −273°C. Conversion from one scale to another is made according to the following relationships:

$$°C = [°F − 32] \times 5/9$$
$$°F = [°C \times 9/5] + 32$$
$$°K = °C + 273$$

As an example, suppose we want to convert 86°F to °C and °K. We proceed as follows:

$$86°F = [86 − 32] \times 5/9 \ °C$$
$$= 54 \times 5/9 \ °C$$
$$= 30°C$$
$$= 30 + 273 = 303°K$$

Temperature is measured with several different instruments, the most common of which is a glass thermometer containing either mercury or alcohol. As temperature rises, the liquid in the tube expands. By special design, these thermometers can also be used to record maximum and minimum temperature over a period of time.

A "thermocouple" is a metal strip composed of two different metals which expand and contract at a different rate as temperature changes. When the instrument is connected to a recording pen moving across a rotating drum it is called a thermograph. A recording instrument such as this provides a continuous record of temperature at a particular location.

Pressure

Like all fluids, the atmosphere has weight. Because the weight of the entire atmosphere does not change significantly with time, its total weight is not an important factor in meteorology. What is important is how the weight of the atmosphere above one point differs from the weight above another point some distance away. Pressure expresses the weight of the atmosphere over a

FIGURE 2.3
Temperature Scales

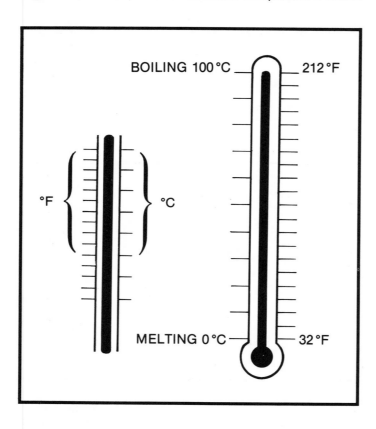

BOILING 100°C — 212°F

°F   °C

MELTING 0°C — 32°F

*Whereas water seldom freezes at 32°F, ice will melt at this temperature.

particular location in terms of weight per unit area. It is commonly expressed in terms of pounds per square inch [lb/in²], which reflects the weight of air over one square inch of surface area. In the metric system, pressure is expressed in terms of dynes* per square centimeter. By convention, meteorologists express pressure in terms of millibars [mb], a unit equivalent to 1000 dynes per square centimeter. Average sea-level pressure in this system of units is 1013 mb which is equivalent to 1,013,000 dynes per square centimeter.

Because of complicated patterns of air motion as well as variable temperature [cold air is heavier than warm air] and water vapor content [moist air is lighter than dry air], pressure at a particular location is continually changing. In this way, horizontal pressure differences are established which control wind direction and speed. In the same way

that air rushes out of an inflated automobile tire when the valve is opened, air rushes away from high pressure centers and toward low pressure centers. Because of the earth's rotation, air does not move directly from high to low pressure but spirals inward counterclockwise toward low pressure areas and outward clockwise from high pressure areas [Figure 2.4].

Thus, the distribution of pressure over the earth's surface controls the direction and speed of wind patterns. Large differences in pressure over small areas produce high-speed winds. An extreme example is a tornado, a storm characterized by very low pressure at its center and rapidly increasing pressure from the center outward producing winds sometimes in excess of 250 miles per hour.

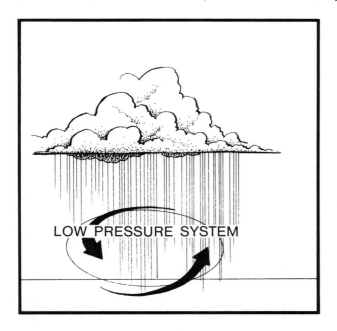

LOW PRESSURE SYSTEM

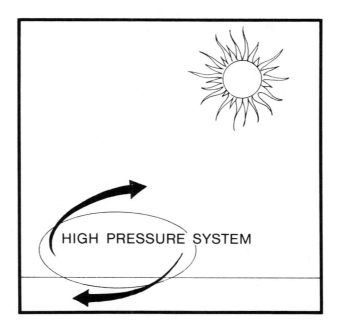

HIGH PRESSURE SYSTEM

FIGURE 2.4 A
Counterclockwise-Convergent Wind Movement in a Northern Hemisphere Low Pressure Center

FIGURE 2.4 B
Clockwise-Divergent Wind Movement in a Northern Hemisphere High Pressure Center

*In the centimeter-gram-second system of units, mass is measured in grams and force or weight is measured in dynes. Average sea-level pressure is 14.7 pounds per square inch which is equivalent to 1,013,000 dynes per square centimeter or 1013 millibars.

FIGURE 2.5
A Mercurial Barometer.
Atmospheric pressure is
proportional to the height
of mercury in the tube.

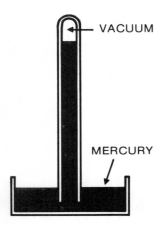

VACUUM

MERCURY

FIGURE 2.6
Schematic Illustration of an
Aneroid Barometer

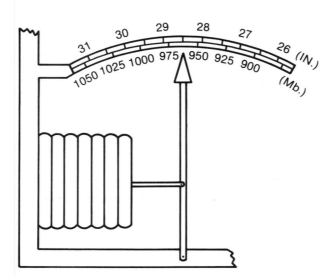

A barometer is an instrument for measuring pressure. Two types are commonly used, the mercurial barometer and the aneroid barometer. The mercurial barometer (Figure 2.5) is a glass tube, sealed at one end, from which air has been evacuated. The open end of the tube is immersed in a reservoir of mercury. As the atmosphere pushes down on the surface of the reservoir, mercury is forced up the tube to a height which is proportional to the atmospheric pressure. Thus, pressure is often expressed in terms of inches of mercury.

Because of its small size and portability, the aneroid barometer (Figure 2.6) is more widely used than the mercurial barometer even though it is not always as accurate. It consists of a sealed metal chamber from which the air has been partially evacuated and which is kept from collapsing by a spring. Changes in atmospheric pressure cause the chamber to expand and contract. These changes in the width of the chamber are magnified and transmitted to an indicator. If a pen is attached to the indicator and placed on a rotating drum, the instrument becomes a barograph and makes a continuous record of atmospheric pressure.

Humidity

Water (H_2O) occurs in the atmosphere as a liquid (clouds and rain), as a solid (snow, sleet, hail), and as a gas (water vapor). Humidity is a measure of water vapor (gas) in the atmosphere. It varies with time and place from near zero to a maximum of 4% (4 parts water vapor to 100 parts air). South and east Texas, particularly near the Gulf Coast, are characterized by high humidity, whereas the air in arid and semi-arid regions of West Texas is low in water vapor content.

Water vapor content of air is usually expressed as "relative humidity," the ratio of the amount of water vapor in the air to the amount possible for a particular temperature. Since warm air can hold more water vapor than cold air, as the temperature increases, so does the capacity of the air for retaining water vapor. If this capacity is exceeded, some of the vapor condenses and forms water droplets. The process is similar to water droplets forming on the outside of a glass of ice water on a hot humid day. The glass cools the air in contact with it until it becomes "saturated," the point at which the air is holding all of the water vapor possible. Further cooling results in condensation, the transition of water from a gas to a liquid, which is the opposite of evaporation. Condensation in the atmosphere produces clouds or fog and, on the earth's surface, dew or frost.

There are two ways to change relative humidity. One is to change the amount of water vapor actually in the air and the other is to change the amount of water vapor the air can hold, which since hot air can hold more than cold air, is accomplished by raising or lowering the temperature. Changing the temperature is by far the most important. As illustrated in the next chapter, cooling the air to its dewpoint temperature (the temperature to which air must be cooled in order for condensation to occur) is the cause of most clouds and precipitation. This strong dependency on temperature is the reason why, on most days, relative humidity is maximum in the early morning when the air is coolest and reaches its minimum during the hot afternoon. This occurs even when the amount of water vapor in the air does not change.

None of the techniques to measure atmospheric humidity is completely satisfactory. The instruments most commonly used are the "hair hygrometer" and the "psychrometer." The hair hygrometer is based on the fact that hair absorbs moisture and expands when humidity is high. Human hair expands by about 2½% as the relative humidity increases from 0 to 100%. The hygrometer registers the change in length of one or more hairs and expresses this in relative humidity units.

The psychrometer [Figure 2.7] consists of a pair of thermometers, one of which has a bulb wrapped with a piece of muslin cloth which has been wetted with pure water. When the thermometers are ventilated, the dry bulb will indicate the air temperature and the wet bulb will be cooled because of the evaporation of water from the bulb [cloth]. The principle is the same as that which controls the heat-balance system of the human body. Humans perspire so that perspiration evaporation cools the skin. Thus, evaporation results in cooling the surface from which the evaporation takes place. How much evaporation occurs depends on the amount of water vapor already in the air. If the air is saturated, no evaporation can occur, and the temperature of the wet bulb does not decrease. This explains why problems with heat exhaustion and heat stroke are more common in humid climates where the cooling mechanism of the human body is inefficient. For very dry air, evaporation from the wet bulb will occur readily and the wet-bulb temperature may be depressed far below the actual air temperature. The difference in actual temperature and wet bulb temperature is called the wet-bulb depression and can be converted to dewpoint temperature and relative humidity by using Tables 2.1 and 2.2. Table 2.1 shows dewpoint temperature as a function of temperature and wet-bulb depression and Table 2.2 shows relative humidity as a function of the same two variables. An example is shown with shading in the table. For a temperature of 60 °F and a wet-bulb temperature of 50 °F, the wet-bulb depression is 10 °F [60° - 50°]. From Table 2.1, for a temperature of

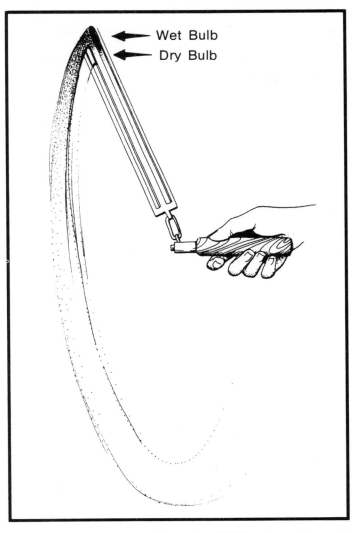

Wet Bulb
Dry Bulb

FIGURE 2.7
Humidity Observation with a Sling Psychrometer

60 °F and a depression of 10 °F, the dewpoint temperature is 40 °F. From Table 2.2, for a temperature of 60 °F and a depression of 10 °F, the relative humidity is 48%. Thus, cooling this air to 40 °F would produce a relative humidity of 100%. Further cooling would produce condensation.

13

TABLE 2.1
Saturation Vapor Pressure
in Inches of Mercury and
Temperature of Dew Point in
Degrees Fahrenheit

[Barometric pressure, 30.00 inches]

AIR TEMP	SATURATION VAPOR PRESSURE	DEPRESSION OF WET-BULB THERMOMETER													
[°F]	[IN.]	1	2	3	4	6	8	10	12	14	16	18	20	25	30
0	.038	−7	−20												
5	.049	−1	−9	−24											
10	.063	5	−2	−10	−27										
15	.081	11	6	0	−9										
20	.103	16	12	8	2	−21									
25	.130	22	19	15	10	−3	−15								
30	.164	27	25	21	18	8	−7								
35	.203	33	30	28	25	17	7	−11							
40	.247	38	35	33	30	25	18	7	−14						
45	.298	43	41	38	36	31	25	18	7	−14					
50	.360	48	46	44	42	37	32	26	18	8	−13				
55	.432	53	51	50	48	43	38	33	27	20	9	−12			
60	.517	58	57	55	53	49	45	40	35	29	21	11	−8		
65	.616	63	62	60	59	55	51	47	42	37	31	24	14		
70	.732	69	67	65	64	61	57	53	49	44	39	33	26	−11	
75	.866	74	72	71	69	66	63	59	55	51	47	42	36	15	
80	1.022	79	77	76	74	72	68	65	62	58	54	50	44	28	−7
85	1.201	84	82	81	80	77	74	71	68	64	61	57	52	39	19
90	1.408	89	87	86	85	82	79	76	73	70	67	63	59	48	32
95	1.645	94	93	91	90	87	85	82	79	76	73	70	66	56	43
100	1.916	99	98	96	95	93	90	87	85	82	79	76	72	63	52

TABLE 2.2
Relative Humidity, Per Cent

DEPRESSION OF WET-BULB THERMOMETER

AIR TEMP [°F]	1	2	3	4	6	8	10	12	14	16	18	20	25	30
0	67	33	1											
5	73	46	20											
10	78	56	34	13										
15	82	64	46	29										
20	85	70	55	40	12									
25	87	74	62	49	25	1								
30	89	78	67	56	36	16								
35	91	81	72	63	45	27	10							
40	92	83	75	68	52	37	22	7						
45	93	86	78	71	57	44	31	18	6					
50	93	87	80	74	61	49	38	27	16	5				
55	94	88	82	76	65	54	48	33	23	14	5			
60	94	89	83	78	68	58	48	39	30	21	13	5		
65	95	90	85	80	70	61	52	44	35	27	20	12		
70	95	90	86	81	72	64	55	48	40	33	25	19	3	
75	96	91	86	82	74	66	58	51	44	37	30	24	9	
80	96	91	87	83	75	68	61	54	47	41	35	29	15	3
85	96	92	88	84	76	70	63	56	50	44	38	32	20	8
90	96	92	89	85	78	71	65	58	52	47	41	36	24	13
95	96	93	89	86	79	72	66	60	54	49	44	38	27	17
100	96	93	89	86	80	73	68	62	56	51	46	41	30	21

Clouds

Clouds are composed of liquid water and/or ice particles suspended in the air. In order to recognize and identify clouds it is necessary to classify them according to appearance and altitude. Proper interpretation of cloud appearance is an important clue in determining atmospheric motion at a given time.

Almost all clouds result from the rapid cooling of air as it moves upward in the atmosphere. "Stratus" clouds are layers or sheets usually covering a large horizontal area and result from the gradual lifting of entire layers of air without strong local upward vertical currents. "Cumulus" clouds have a billowy or heaped-up appearance and are characterized by large vertical development due to strong localized upward motion of up to 30 or more centimeters per second. High-level clouds, usually above 22,000 feet [7 kilometers], are composed entirely of ice crystals and are called "cirrus" clouds. They may exhibit significant vertical or horizontal development in which case they are referred to as either "cirrocumulus" or "cirrostratus." Similarly, middle-level clouds at around 7500 - 22,000 feet [2 1/3 - 7 kilometers] are designated by the prefix "alto." Therefore, "altocumulus" are middle-level clouds with vertical development and "altostratus" are those with horizontal development.

The word "nimbus" is used as a prefix or suffix to indicate clouds producing precipitation. "Cumulonimbus" clouds are clouds which may produce severe weather such as thunderstorms, hail, heavy rain and tornadoes. "Nimbostratus" are layered clouds from which light or moderate rain or snow usually occurs. "Fog" is merely a cloud at ground level. When fog is formed of tiny ice crystals rather than water droplets, it is called an ice fog. Table 2.3 provides a summary of basic cloud types.

Precipitation

When water [or ice] droplets in a cloud grow large enough to fall under the influence of gravity, precipitation results. Precipitation can be divided into three basic classes: liquid, solid, and freezing. The fundamental types of liquid precipitation are "rain" and "drizzle," the difference being the size of the droplets. The diameter of drizzle is generally less than 0.5 mm. The two types of solid precipitation are "snow" and "hail." Snow develops when ice crystals form within a cloud and fall to the ground through cold air so that the flakes do not melt. Hail is small balls or chunks of ice with diameters ranging from 0.2 to 2 inches [5mm to 50mm] or more that fall from cumulonimbus clouds during thunderstorms. These often destructive stones are produced by the freezing of successive layers of water as the ice moves through a thick cloud layer.

Freezing precipitation is of two types. "Sleet" is small ice particles or pellets that originate as rain but freeze as they fall through a sub-freezing layer of air between the cloud and the ground. "Freezing rain or drizzle" forms when rain [drizzle] freezes on impact with the ground or other object. Figure 2.8 illustrates the formation of sleet, rain and snow.

The amount of precipitation is measured by a rain gauge. The standard rain gauge is cylindrical and exactly 8 inches [20 centimeters] in diameter, although any open vessel of the same cross section throughout will suffice. The depth is measured by a small rule in inches or millimeters. Snow, sleet and hail are melted and the equivalent liquid depth recorded. The ratio of snow depth to liquid equivalent is usually about 10 to 1 and is occasionally as much as 30 to 1.

There are several rain gauges which provide a continuous record of precipitation and determine the rate of rainfall over short periods of time. The "tipping bucket gauge" can be used only for rain, whereas the "weighing rain gauge" can be used for either rain or snow. In both devices, rainfall is recorded by a pen on a revolving drum.

Rain gauges should be placed in the open away from buildings or trees which might shield the gauge from precipitation. The top of the gauge must always be level.

Wind

Wind is air in motion relative to the earth's surface.

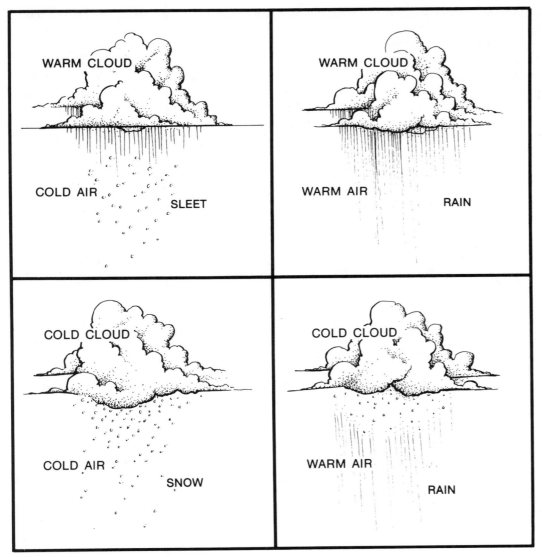

FIGURE 2.8
Precipitation Types

CLOUD TYPE	DESIGNATION	TYPICAL HEIGHT IN MID-LATITUDES
High clouds		5-13 km
Cirrus	Ci	[16,500-
Cirrocumulus	Cc	45,000 feet]
Cirrostratus	Cs	
Middle clouds		2-7 km
Altocumulus	Ac	[6,500-
Altostratus	As	23,000 feet]
Low clouds		Below 2 km
Stratus	St	[6,500 feet]
Stratocumulus	Sc	
Nimbostratus	Ns	
Vertically Developed		Cloud base at low levels with
Cumulus	Cu	considerable vertical growth
Cumulonimbus	Cb	Cb clouds may reach 18.5 km [60,000 feet].

TABLE 2.3
Classification of Cloud Types

17

Although air moves vertically as well as horizontally, the speed of vertical motion is usually much less than horizontal motion. Nevertheless, vertical motion is principally responsible for the formation of clouds and precipitation. Only the horizontal motion is measured on a regular basis, however, and it is this that is referred to as wind.

In order to describe the wind field at a particular time, it is necessary to specify both the speed and direction of motion. Wind direction is expressed as the direction from which the wind is blowing. A south wind blows from the south toward the north and a northwest wind blows from northwest to southeast. Direction is also measured and described in degrees from north as shown in Figure 2.9. North is assigned a direction of 0° (or 360°). The number of degrees increases clockwise until a complete circle of 360° has been completed. A 45° wind is a northeast wind, a 180° wind is a south wind, a 315° wind a northwest wind, and so on around the compass.

Wind vanes (Figure 2.10) have been used since ancient times as indicators of wind direction. In order to obtain a continuous record of wind direction, the vane may be connected by electrical circuits to a pen which records a trace on a rotating drum. There are also indicators which show the instantaneous wind direction by means of a pointer on a dial.

Wind speed is expressed in either miles per hour, nautical miles per hour (knots), or meters per second. One knot is equivalent to 1.15 miles per hour. The National Weather Service uses knots to express both surface and upper-level winds. Many instruments are used to measure wind speed. Surface wind speeds are measured with anemometers. The most common is the cup anemometer which consists of three or more hemispherical cups on a vertical shaft (Figure 2.11). The number of rotations of the cups in a specified time is a measure of wind speed and may be converted to any of the units commonly used. The direction and speed of upper-level winds are determined by tracking an ascending gas-filled balloon from the surface up through the atmosphere. Periodic readings are taken of the balloon's position which reveal average wind direction and speed between balloon positions. More refined systems include a "rawinsonde," a self-tracking radio direction-finding unit that measures the position of the balloon and its distance from the observing station.

FIGURE 2.9
Wind Direction in Degrees

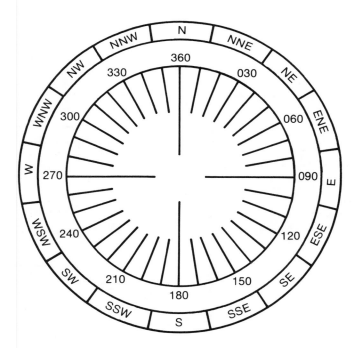

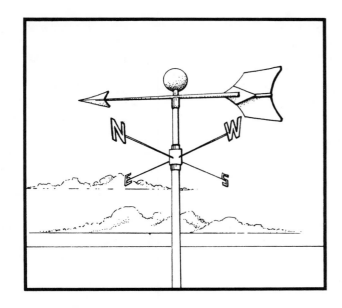

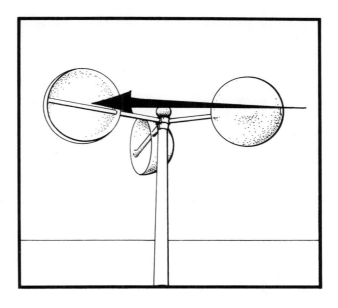

FIGURE 2.10
Wind Vane for Measuring
Wind Direction

FIGURE 2.11
Cup Anemometer for
Measuring Wind Speed

Weather Maps

A weather map provides a summary of concurrent weather events over a large area. Observations of temperature, pressure, dewpoint temperature, wind speed and direction, clouds and present weather are taken all over the world, exchanged internationally, and then collected at central locations.

So that the state of the atmosphere can be readily interpreted at any individual station, data are plotted on weather maps according to a standard format. An abbreviated station model showing the plotting positions for most observations is shown in Figure 2.12a. An example utilizing actual data is shown in Figure 2.12b.

The example illustrates a station where the temperature and dewpoint temperature are 52°F and 38°F, respectively. Wind direction is north (from north to south) as indicated by the direction the staff is pointing. Attached to the staff are barbs indicating wind speed; a full barb is 10 knots and a half-barb is 5 knots. Wind

speed in the example is 15 knots. Cloud cover is indicated by a progressive darkening of the station circle. If no shading is shown, skies are clear; partial shading indicates partly cloudy skies. Here, the station circle is filled in completely indicating overcast skies. All station pressures reflect the pressure which would occur if the station were at sea level. This eliminates pressure differences due to differences in station elevation and allows a comparison of pressure changes at various locations due to changing weather patterns. Pressure is decoded by adding a decimal point before the last digit and preceding the indicator with ten.* Thus, 122 is decoded as 1012.2 mb. Other plotted observations indicate that it was raining at the time of the observation [· ·] and that pressure was rising continuously [/] and had risen 1.2 mb during the past three hours. A summary of plotting symbols for clouds and weather is given in Table 2.4

Information is also collected, coded and plotted on maps to show upper-atmosphere conditions. A plotting model

*For sea-level pressures below 1,000 mb, the indicator is preceded by 9 rather than 10. Therefore, 982 indicates a pressure of 998.2 mb.

for upper-air observations is illustrated in Appendix 4.

After observations are plotted on the appropriate map, the map is analyzed to produce a pictorial three-dimensional description of the atmosphere. Each map represents the air at a particular level. The surface map is the most complete, both in the number of stations reporting and the number of atmospheric properties measured. The surface map is the familiar weather map that appears in simplified form in newspapers and television weather reports.

FIGURE 2.12
Abbreviated Station Model

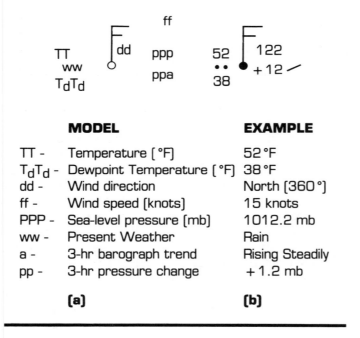

	MODEL	EXAMPLE
TT -	Temperature [°F]	52°F
T_dT_d -	Dewpoint Temperature [°F]	38°F
dd -	Wind direction	North [360°]
ff -	Wind speed [knots]	15 knots
PPP -	Sea-level pressure [mb]	1012.2 mb
ww -	Present Weather	Rain
a -	3-hr barograph trend	Rising Steadily
pp -	3-hr pressure change	+ 1.2 mb

(a)   **(b)**

Figure 2.13 shows a simplified surface weather map with data and analysis-lines omitted. The map is characterized by two major low-pressure systems, one centered in eastern Illinois and the other approaching the northwest Pacific coast. These "lows" or "cyclones" are moving from west to east producing characteristic weather along their paths. As indicated on the map, the pressure between these two systems is relatively high. Arrows show the predominant wind flow to be counterclockwise and converging [moving toward low pressure] in the vicinity of each low. The wind pattern in association with the "high" is clockwise and diverging [moving away from high pressure].

Air masses with different characteristics are distinguished by "fronts," shown as heavy lines in Figure 2.13. A front is a narrow transition zone across which temperature, humidity, and wind direction change abruptly. The cold front is a line along which cold air is replacing warm air at the surface and is identified by a series of triangles on the side of the line toward which the air is moving. The warm front is similarly indicated by a series of half circles. Note the temperature contrasts shown on either side of the fronts in Figure 2.13.

The rain and snow indicated on the map may be related to a combination of causes. Recall that low pressure centers are areas of low-level horizontal convergence which produce upward vertical motion in much the same way that toothpaste emerges from a tube when it is squeezed. These centers may create clouds and precipitation. Also, cold air is heavier than warm air, producing upward motion of warm air in the vicinity of fronts as the cold air wedges underneath. The relationship between upward vertical motion and clouds and weather will be explored in more detail in Chapter 3.

TABLE 2.4
Plotting Symbols for
Clouds and Weather

CLOUDS

LOW CLOUDS

⌒ Cu
△ Cu with development
🜊 Cu without clear tops
⊶ Sc from Cu
∼ Sc
— St
⋯ Scud
⏢ Cu and Sc
🜊 Cb

MIDDLE CLOUDS

∠ As thin
◢ As thick
∼ Ac thin
∠ Ac thin-changing
⟋ Ac thin bands
⋊ Ac from Cu
◢ Ac layered
M As tufts, turrets
◢ Ac mixed

HIGH CLOUDS

→ Ci
⇁ Ci dense
⇁ Ci dense from Cb
⟋ Ci
⌐ Ci and Cs
⟋ Ci and Cs
⊷ Cs veil
→ Cs
∠ Cc

PRESENT WEATHER

' Drizzle
• Rain
✻ Snow
⟶ Ice needles
△ Granular snow
⟶✻ Snow crystals
△ Ice pellets
⟨ Lightning
ℝ Thunder

ℝ̇ Thunderstorm with rain
ℝ Heavy thunderstorm
)(Funnel cloud
∿ Freezing drizzle
∿ Freezing rain
= Fog
∞ Haze
⌒⌒ Smoke
S Dust

$ Blowing dust
⊢ Low drifting snow
⊢ High drifting snow
⊷ Precipitation not reaching ground
)•(Precipitation distant
(•) Precipitation near station
▽ Rain shower
△̱ Hail
✻̱ Snowshower
▽

[x] means x occurred within the last hour.

Heavier precipitation or more intense activity is indicated by a modification of the symbols;

example: '' light,   ''' moderate,   '·' heavy drizzle

Combinations may occur: •✻ rain mixed with snow   ℝ̂ severe thunderstorm with hail

FIGURE 2.13
A Simplified Surface Weather
Map Showing Fronts, Wind
Direction, Temperature,
Pressure Centers and
Precipitation

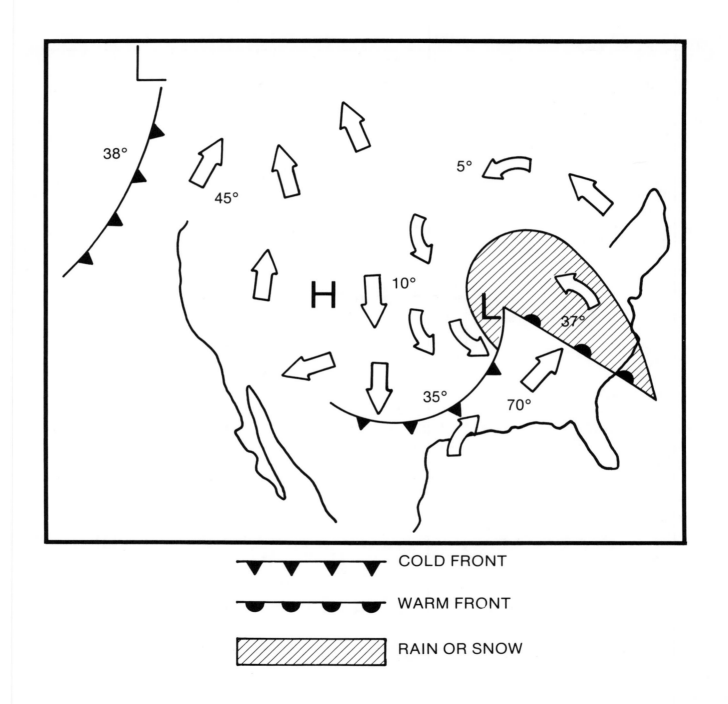

▼▼▼ COLD FRONT

●●● WARM FRONT

▨ RAIN OR SNOW

The energy which causes atmospheric motion is provided by the sun. Incoming solar radiation heats the earth but not uniformly. The heat input at the equator is significantly greater than that at the poles. Because the earth is tilted on its axis of rotation, the heating also varies with the seasons. Figure 3.1 illustrates the orientation of the earth and sun during winter, spring, summer and fall.

Note that as the earth revolves around the sun (once in 365¼ days), the portion of the earth's surface which is directed toward the sun changes. In positions 1 and 3, the

ATMOSPHERIC CIRCULATION

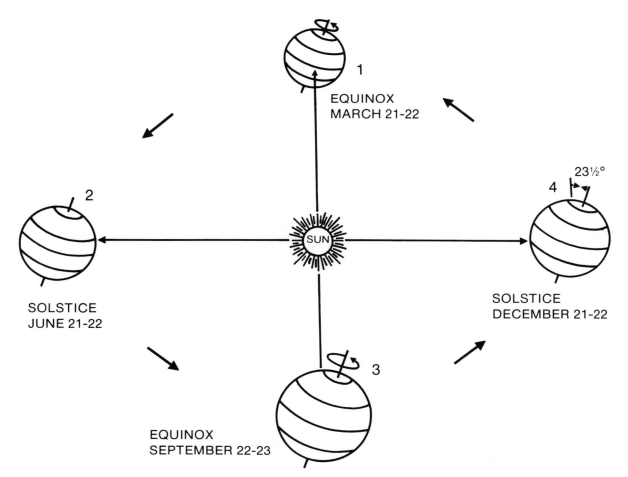

FIGURE 3.1
Orientation of the Earth with Respect to the Sun at Different Times of the Year

sun is directly overhead at the equator, day and night are 12 hours long over the entire earth and equal amounts of energy are distributed over each hemisphere. In position 4 the northern hemisphere is tilted away from the sun while the southern hemisphere is receiving more of the direct radiation. Thus, the northern hemisphere is experiencing winter while it is summer in the southern hemisphere. This situation is reversed six months later when the northern hemisphere is pointed toward the sun and the southern hemisphere is tilted away as illustrated by position 2.

Because of the unequal solar heating, spatial temperature differences develop, with maximum temperatures occurring near the equator and minimum temperatures at the poles. Since warm air is lighter than cold air, this equator-to-pole temperature difference causes a simple circulation to develop as illustrated in Figure 2.1 in Chapter Two. Warm air rises at the equator and is transported at upper levels toward the poles. The cold air sinks at the poles and moves toward the equator at low levels.

Unfortunately, this simple description of wind patterns does not tell the complete story. The interaction of many factors whose influence is not completely understood produces a much more complicated circulation pattern shown in Figure 3.2.

Note that the basic features which characterize the circulation have not changed: warm air is rising at the equator and cold air is sinking at the poles. As the warm air rises, however, it cools [for reasons which will be discussed in the next section] and begins to sink at subtropical latitudes [20° to 30° N] long before it reaches the pole. This sinking air then spreads both north and south when it reaches the earth's surface. The air moving from south to north eventually encounters the southerly moving air which has resulted from the sinking of cold air at the poles. Because warm air is lighter than cold air, the warm air is forced to rise at this boundary known as the polar front.

At this point, another important factor, called the "Coriolis effect," must be introduced. Imagine a missile being launched from the north pole on a trajectory toward the equator [Figure 3.3]. If the travel time of the missile from pole to equator was approximately an hour, the earth would have rotated 15 degrees to the east while the missile was in flight. To an earthbound observer it would appear that the missile had veered off course and hit the earth at a point 15 degrees west of its target. However, someone in space observing the earth would see an accurate picture. The trajectory of the missile was a straight line. It was the rotation of the earth beneath the missile which produced the apparent deflection. Thus, in the northern hemisphere, the rotation of the earth pro-

FIGURE 3.2
Three-cell Circulation Model of Simplified Atmospheric Motion

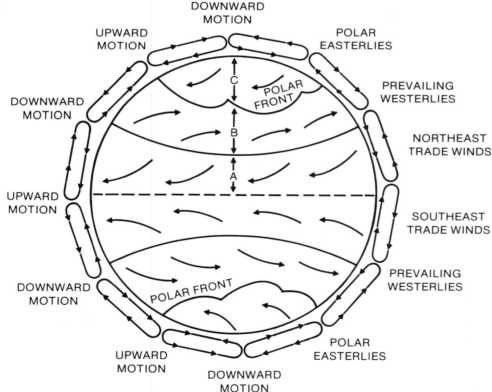

duces a deflection to the right of the path of a free-moving object, including air. The deflection in the southern hemisphere is to the left. This Coriolis effect—or change in wind direction due to the rotation of the earth—is extremely important in understanding atmospheric circulation patterns.

If the earth were not rotating, wind direction would be determined primarily by the temperature difference between the equator and pole. Prevailing winds in the three zones shown in Figure 3.2 for the northern hemisphere would be northerly in Zone A, southerly in Zone B and northerly in Zone C. On a rotating earth, the Coriolis effect causes a deflection to the right resulting in the wind patterns shown. In Zone A, weather features move from east to west in a wind regime known as the "trade winds." In middle latitudes, weather systems move from west to east in the "prevailing westerlies." In polar regions, the circulation is once again east to west and designated the "polar easterlies." That most weather systems affecting Texas (Zone B) move from west to east is common knowledge to most Texans. Exceptions are tropical storms and hurricanes which may approach the Texas coast from the east or southeast in Zone A during late summer and fall. When these storms travel far enough north to enter the region of the prevailing westerlies they abruptly change direction and proceed from west to east. This curved path is typical of most hurricanes. They enter from east to west in the band of the trade winds and depart from west to east as they are caught up in the prevailing westerlies to the north. This pattern will be discussed in Chapter 4.

Therefore, weather systems have a preferred direction of motion which depends upon latitude. However, the Texas wind doesn't always blow from west to east, of

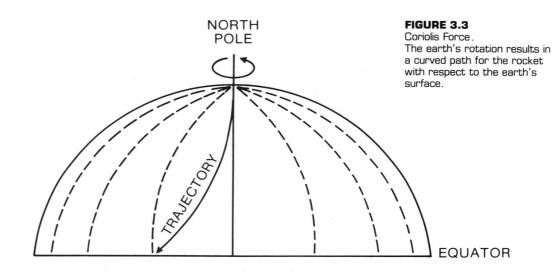

FIGURE 3.3
Coriolis Force.
The earth's rotation results in a curved path for the rocket with respect to the earth's surface.

course. This is due to smaller scale wind patterns that occur in association with low pressure centers (cyclones) and high pressure centers (anticyclones) which develop within the circulation. In middle latitudes the cyclones (and anticyclones) do move from west to east in the prevailing westerlies, but they carry with them a circulation of their own which was described briefly in Chapter 2 and illustrated in Figures 2.4a and 2.4b. As air rushes in toward low pressure centers, the Coriolis effect causes it to be deflected to the right resulting in the counter-clockwise circulation shown in Figure 2.4a. As air rushes away from high pressure centers, the Coriolis effect produces a clockwise circulation. Thus, air spirals inward (convergence) counterclockwise in low pressure areas and outward (divergence) clockwise in high pressure areas.

VERTICAL MOTION, CLOUDS AND WEATHER

A comparison of those regions which experience upward vertical motion with areas of heavy precipitation suggests a strong correlation between the two. Note from Figure 3.2 that mean upward vertical motion occurs at the equator and at the polar front with dominant downward motion in the subtropics and at the pole. Equatorial regions, where motion is upward, receive abundant precipitation throughout the year. Heavy precipitation falls also along the polar front where warm air is forced to move upward by colder, heavier air moving from the poles. On the other hand, many of the major deserts of the world are found in the subtropics where motion is predominantly downward. Precipitation is also light or nonexistent in polar regions which are characterized by subsiding air. Practically all clouds and weather producing systems owe their existence to upward motion; in order to produce weather it is first necessary to force air to rise.

Clouds are produced by condensation of water vapor in the atmosphere. All air contains water vapor to some extent but recall from Chapter 2 that warm air can hold more vapor than cold air. When air is cooled, its capacity for retaining water vapor is reduced. When the temperature is lowered to the dewpoint, water vapor capacity is reached and further cooling results in condensation, producing clouds and perhaps precipitation.

But how does lifting the air (upward vertical motion) produce cooling? The answer explains the existence of essentially all weather producing systems and deserves careful consideration. When a gas expands, it cools. This is demonstrated by a simple experiment. With your mouth wide open and your hand approximately two inches away, blow on your hand. The air should be warm as it leaves your mouth at body temperature. Now repeat the experiment with your lips puckered tightly. As the air exits the restricted opening, it is forced to rapidly expand into the atmosphere. This air feels considerably cooler even though your hand is still held about two inches from your mouth. The air has been cooled by rapid expansion.

Similar expansion occurs when air rises in the atmosphere. Since pressure, which represents the weight of the air above you, always decreases with altitude, a rising mass of air will expand as it encounters lower pressure. As the air rises and expands, it cools in exactly the same way the air exiting your mouth cooled. If the air rises sufficiently to cause cooling to the dewpoint, condensation begins and a cloud develops. Depending on the magnitude of the upward motion, some clouds will grow to a size sufficient to produce precipitation. All clouds, with the exception of fog, are produced in this way as rising air expands and cools. Similarly, descending motion warms the air and is related to the absence of clouds and precipitation.

TRIGGERING UPWARD VERTICAL MOTION

There are several ways to initiate upward vertical motion and produce weather. Each is important and related to a particular type of weather system. These "triggering mechanisms" can be separated into four groups:

1. Upward vertical motion due to convergence.
2. Upward vertical motion due to a front.
3. Upward vertical motion due to convection.
4. Upward vertical motion due to orographic lifting.

Upward Motion Due to Convergence

As illustrated in Figures 2.4a and 2.4b, the circulation around low pressure areas (cyclones) in the northern hemisphere is counterclockwise and convergent while that around high pressure areas (anticyclones) is clockwise and divergent. Figure 3.4 illustrates the variation of wind direction in each of these systems and the vertical motion which results. In the cyclone, the air spirals inward counterclockwise while in the anticyclone the air spirals outward clockwise.

In much the same way that toothpaste is squeezed from the top of the tube, air in a cyclone is forced upward by low level convergence. The inward motion effectively squeezes the air horizontally which forces it to stretch vertically. Since the air cannot go downward into the ground, it moves upward into the path of least resistance. In the case of high pressure, the air diverging near the surface must be replaced by descending air from above; therefore, the vertical motion is downward [Figure 3.4].

As discussed previously, rising air expands and cools. If cooled to the dewpoint temperature, condensation occurs producing a cloud. Subsequent growth of the cloud may then lead to precipitation. Because of low level convergence and the resultant upward motion, low pressure areas—provided atmospheric moisture is available—are preferred regions for clouds and precipitation. In high pressure areas, motion is downward which warms the air due to compression and discourages weather development.

The strong correlation between pressure and weather patterns is evident on most barometers designed for average consumers. Rather than noting pressure values, the instrument simply indicates "stormy" at the low-pressure end, "fair" at the high-pressure end, and "changing" in between.

Upward Vertical Motion Due to Frontal Lifting

An air mass is a large body of air that has assumed the characteristics of a particular area of the world where it has resided for a period of time. For example, an air mass moving out of central Canada in the winter tends to be cold and dry while an air mass moving out of the Gulf of Mexico in summer is warm and moist. A simple scheme has been devised to classify air masses according to temperature and moisture content. A cold air mass is designated "polar" and indicated by a P. A warm air mass is "tropical" and indicated with a T. On the basis of

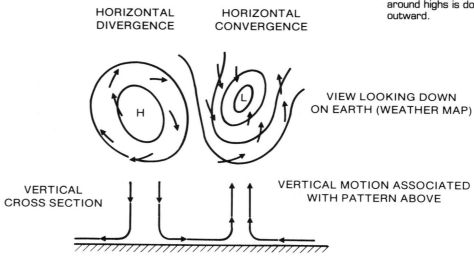

FIGURE 3.4
Convergence and Divergence. The flow around lows is inward and upward while that around highs is downward and outward.

HORIZONTAL DIVERGENCE

HORIZONTAL CONVERGENCE

VIEW LOOKING DOWN ON EARTH (WEATHER MAP)

VERTICAL CROSS SECTION

VERTICAL MOTION ASSOCIATED WITH PATTERN ABOVE

moisture content, an air mass is either "continental" [c] or "maritime" [m]. Consequently, there are four fundamental air-mass types:

mP - maritime polar, cold and moist
mT - maritime tropical, warm and moist
cP - continental polar, cold and dry
cT - continental tropical, warm and dry

Source regions for these air masses in North America are shown in Figure 3.5. The primary source region for maritime polar air is the North Pacific because of the west-to-east progression of weather events in middle latitudes. The east coast of the United States is also affected locally by maritime air from the North Atlantic. The source for maritime tropical air is the Gulf of Mexico and occasionally the Pacific ocean. Continental polar air comes from Canada while continental tropical air masses have their source region in Mexico and the desert southwest of the United States.

FIGURE 3.5
Air Mass Source Regions

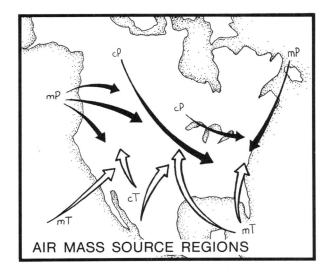

AIR MASS SOURCE REGIONS

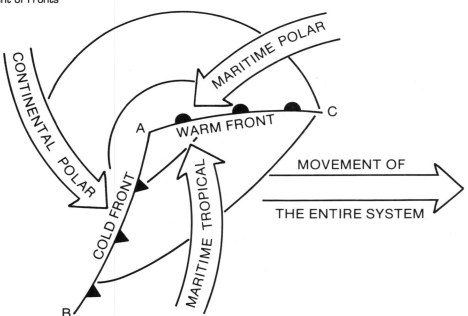

A "front" is a transition zone or boundary between air masses which have been brought together by the wind. Low pressure areas are areas in which air masses from different locations are being brought together and are often associated with the development of fronts. Consider the cyclone shown in Figure 3.6. The counterclockwise circulation and convergence are pulling continental polar air down behind the low as it moves from west to east and, at the same time, pulling maritime tropical air up ahead of the low. Maritime polar air is shown drifting in from the northeast. The boundaries along which these air masses confront one another are fronts. If cold air is replacing warm air at the ground, the boundary is a cold front, while warm air replacing cold air indicates a warm front. The cold front is denoted by the line AB along which continental polar air is replacing maritime tropical air at the ground. The warm front [line AC] is a boundary along which the maritime tropical air from the south is replacing maritime polar air as the system rotates counterclockwise.

When two air masses meet in this way, they do not mix. Since the warm air is lighter than the cold air, it is forced to rise at the frontal boundary. As the warm air rises, it expands and cools leading to the development of clouds and precipitation. Thus, frontal lifting is frequently combined with convergence to produce cloudiness and precipitation in a cyclone.

Examples of frontal lifting are shown in Figure 3.7. A warm front is illustrated in 3.7a and a cold front in 3.7b. Warm fronts, because of the gradual slope of the frontal surface, typically produce steady rain over an extensive area. Cold fronts, with characteristically steeper and faster moving surfaces, frequently produce intense rainfall from cumulonimbus clouds. This rainfall is usually more scattered and of shorter duration than that produced by a warm front.

FIGURE 3.7a
Cross-Sectional View of a
Warm Front

WARM AIR MASS

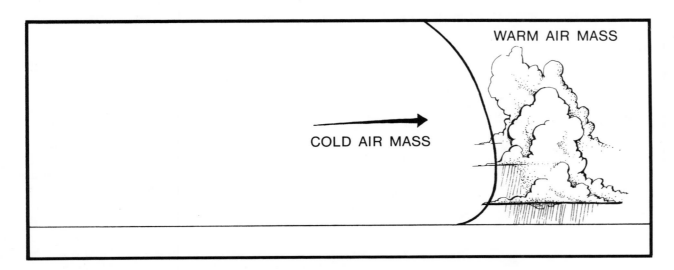

FIGURE 3.7b
Cross-Sectional View of a
Cold Front

WARM AIR MASS

COLD AIR MASS

FIGURE 3.8
Thermal Lifting
Due to Surface
Heating

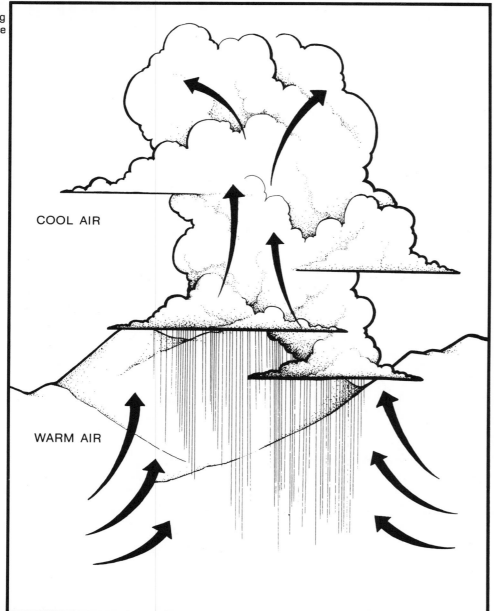

COOL AIR

WARM AIR

Thermal Lifting Due to Small-Scale Convection

On sunny days, convection currents often develop as strong surface heating causes the air near the ground to become bouyant and rise. This "thermal lifting" is most efficient over small areas where heating is especially intense. Individual rising currents may produce scattered cumulus clouds (Figure 3.8). In between the clouds, currents of sinking air replace the surface air that has been lifted by the convection. Because of the alternate rising and sinking currents, the air is very bumpy for aircraft. These convection currents generally dissipate after sunset when solar heating ceases.

Thermal lifting is most pronounced in the warm season and often turns morning stratus clouds into stratocumulus with the possibility of light showers. Continued heating may lead to the development of cumulonimbus clouds resulting in heavier showers and thunderstorms.

Upward Motion Due to Orographic Lifting

When horizontally-moving air encounters hills and mountain ranges, some of the air is lifted which produces expansional cooling and possible clouds and precipitation. As the air passes over the higher land surface, it tends to sink to lower elevations on the opposite side. The side of the topographic feature from which the wind is blowing is called the "windward" side, while the opposite side is referred to as the "leeward" side. Figure 3.9 illustrates how air rises on the windward side of a mountain range. Because of the strong correlation between upward vertical motion and weather as a result of cooling, the windward side of a mountain range is likely to have a wetter climate than the leeward side. The resulting precipitation is commonly called orographic precipitation. The leeward side of a mountain range is generally characterized by a dry climate as a result of the prevailing downward vertical motion. This region is often said to be in the "rain shadow" of the mountains.

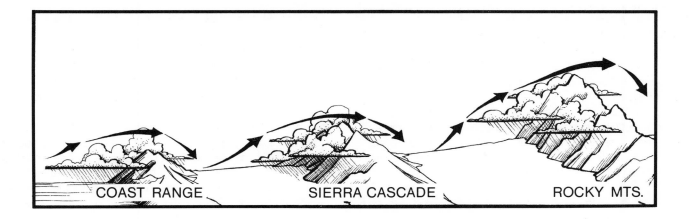

FIGURE 3.9
Orographic Lifting Over
Mountains

COAST RANGE   SIERRA CASCADE   ROCKY MTS.

The following summarize the previous sections of this chapter:

1. Because of the tilt of the earth on its axis as it revolves around the sun, solar heating varies over the earth and with the seasons.

2. The difference in air temperature which results from this unbalanced heating causes the wind.

3. The rotation of the earth complicates wind patterns by twisting them into spirals known as cyclones (low-pressure areas) and anticyclones (high-pressure areas).

4. The distribution of cyclones and anticyclones creates areas of convergence and divergence.

5. Air converging at the surface into lows is forced upward. Air diverging from highs at the surface creates downward motion.

6. Because pressure in the atmosphere always decreases with height, rising air expands and descending air is compressed.

7. When a gas expands, it cools. Compressed air becomes warmer; therefore, rising air cools and descending air warms.

8. Since cold air holds less water vapor than warm air and since rising air cools, clouds and precipitation may result from upward vertical motion.

9. Upward vertical motion is produced by convergence, frontal lifting, orographic lifting and thermal lifting.

Now let's examine specific weather occurrences illustrated by weather maps and identify the mechanisms responsible for producing weather. Shading on each map indicates precipitation. Cold fronts are identified by lines with a series of triangles on the side of the line toward which the front is moving. Warm fronts are similarly identified with half circles rather than triangles. Other solid lines on the chart are lines of constant pressure called isobars. Each pressure center is labeled "low" or "high."

Figure 3.10 shows the weather map for July 3, 1966. The prominent weather feature during the month was an extended heat wave and drought which began in late June. These conditions prevailed from New Mexico to New England because of the dominant high pressure shown on the map. Note that, with the exception of the northwest

WEATHER AND THE WEATHER MAP

corner, high pressure dominates the entire country. This results in downward vertical motion which warms the air and prevents the development of clouds and precipitation.

Figure 3.11 shows conditions at 7 a.m., May 27, 1978. The stationary front extends through the Great Plains, combined with warm, moist air to the east of the front to produce tornadoes, hail and heavy rain. Later on the same day, a tornado was spotted at Edmundson, Texas, near Plainview. A sequence of pictures documenting the entire life cycle of this storm is shown in Figure 4.8. On the previous day, flooding east of Canyon and in Palo Duro Canyon was produced by rains in excess of 10 inches that fell in a period of 90 minutes. Four persons were killed and fifteen injured during the flooding. Property damage was extensive. Conditions at the time of the map were not as severe with rain falling at only a few stations along the front. Note that another front with an associated precipitation pattern is entering the country along the Washington and British Columbia coast.

Conditions for Saturday, August 9, 1980, are illustrated in Figure 3.12. The most significant feature on this map is Hurricane Allen which is approaching the Texas coast. Heavy rain is already occurring along the southern coast ahead of the storm. The eye of the storm moved ashore the next day. Severe damage occurred as a result of high wind, flooding, and tornadoes which formed in association with the storm. A satellite view of Hurricane Allen is shown in Figure 4.10.

FIGURE 3.10

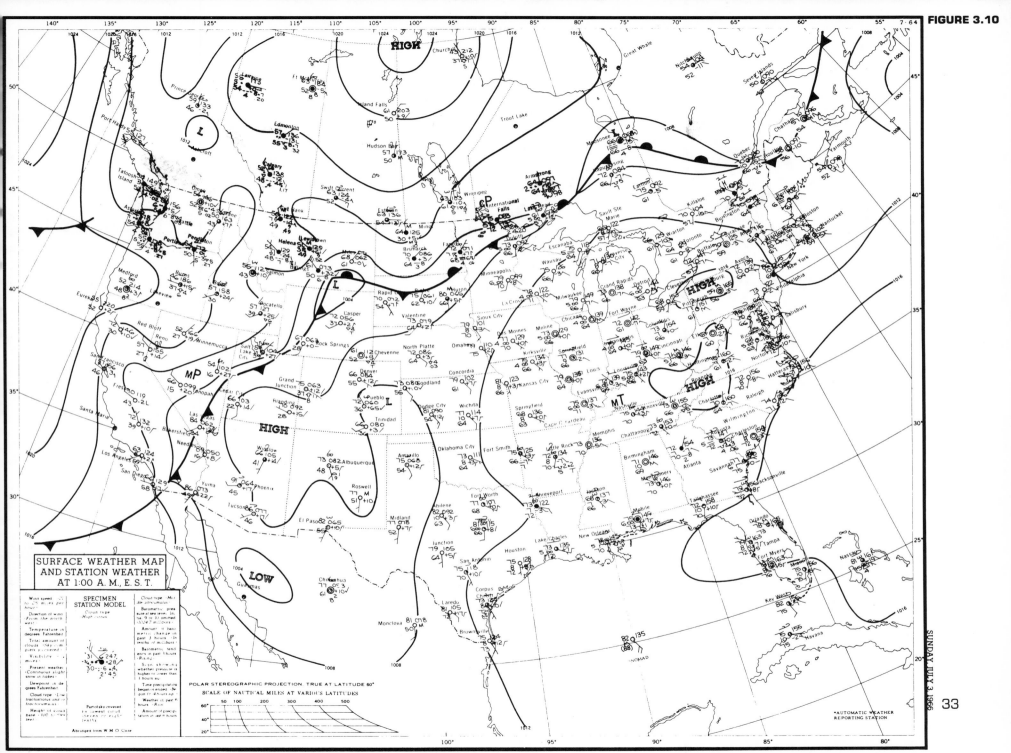

SURFACE WEATHER MAP
AND STATION WEATHER
AT 1:00 A.M., E.S.T.

SPECIMEN
STATION MODEL

POLAR STEREOGRAPHIC PROJECTION. TRUE AT LATITUDE 60°

SCALE OF NAUTICAL MILES AT VARIOUS LATITUDES

+AUTOMATIC WEATHER
REPORTING STATION

Courtesy of National Oceanographic and Atmospheric Administration.

FIGURE 3.11

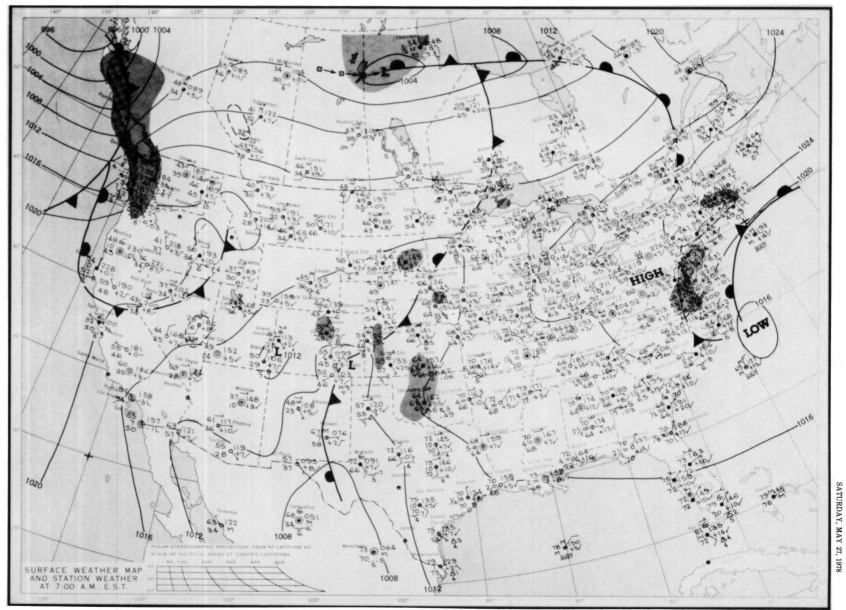

SURFACE WEATHER MAP
AND STATION WEATHER
AT 7:00 A.M. E.S.T.

SATURDAY, MAY 27, 1978

Courtesy of National Oceanographic and Atmospheric Administration.

FIGURE 3.12

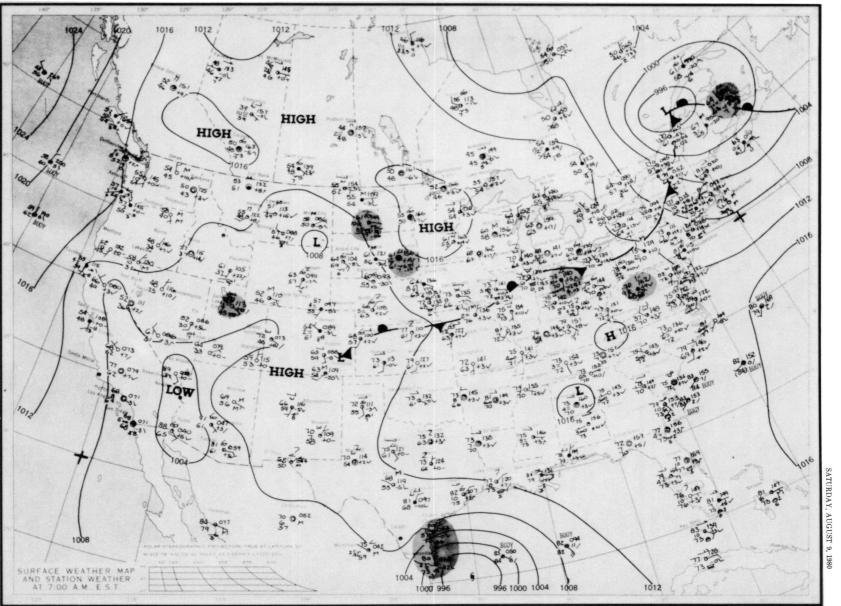

SURFACE WEATHER MAP
AND STATION WEATHER
AT 7:00 A.M. E.S.T.

SATURDAY, AUGUST 9, 1980

Courtesy of National Oceanographic and Atmospheric Administration.

35

Chapter 4 / TEXAS WEATHER

Texas' latitude places it in the zone of the prevailing westerlies [Figure 3.2], the battleground between tropical air masses to the south and polar air masses to the north. Though the climate of Texas is predominantly subtropical, polar air masses invade frequently in winter producing seasonal temperature differences which are most pronounced in the north but are characteristic of the entire state. Even the Rio Grande Valley in the southernmost part of the state occasionally experiences freezing temperatures during winter.

The variability of Texas weather reflects both the size of the state and its mid-latitude location. Its ten climatic regions are illustrated in Figure 4.1. While the eastern half of the state is dominated by abundant moisture from the Gulf of Mexico, the western half becomes progressively semiarid and finally arid as distance from the Gulf increases in a westerly and northwesterly direction. The terrain slopes upward from sea level along the coast to more than 4,000 feet along the Texas-New Mexico border, producing a natural lifting of air which enters the state from the south and southeast. Air from the southwest, on the other hand, is dry. It is this air that dominates the climate of West Texas.

Most of the state is characterized by relatively flat terrain. The mountains in the Trans-Pecos Region are of modest size with fewer than 100 peaks extending to one mile or higher. The tallest is Guadalupe Peak at 8,751 feet. Orographic lifting as a result of these terrain features produces afternoon thunderstorms in the summer which account for most of the annual rainfall in this area.

Maximum rainfall occurs over most of Texas in late spring as a result of thunderstorms. Exceptions are the mountainous Trans-Pecos and the northern High Plains regions which have a summer maximum, and the Pecos Valley, southern Texas and the upper coast which have peak rainfall in September due to tropical storms and occasional hurricanes. Dry summers are the rule in the central portions of the state while dry winters dominate the north and west. In East Texas and along the upper coast, rainfall is distributed throughout the year.

A variety of storm events contribute to Texas weather patterns. Abrupt lifting of air masses along fronts and the lifting due to general convergence in low pressure areas may occur during both warm and cold seasons. During spring and summer this often results in thunderstorms, hail and, in the most severe cases, tornadoes. Warm lows entering the state from the south and southeast may develop into tropical storms and occasionally into hurricanes with heavy rain and damaging winds. Air mass showers and thunderstorms are produced over much of the state during the warm season as a result of thermal lifting or orographic lifting as in the western mountains. When the atmosphere is sufficiently unstable, these storms may develop to severe proportions producing hail and, in some instances, tornadoes.

The remainder of this chapter is devoted to Texas storms and the weather which results.

WEATHER PROFILE

FIGURE 4.1
Texas' Ten Climatic Regions

HIGH PLAINS

LOW ROLLING PLAINS

NORTH CENTRAL

EAST

TRANS—PECOS

EDWARDS PLATEAU

SOUTH CENTRAL

UPPER COAST

SOUTHERN

LOWER VALLEY

FIGURE 4.1
Texas' Ten Climatic Regions

As described in Chapter 3, fronts form in association with low pressure systems and represent boundaries separating air masses with contrasting temperatures and/or water vapor content. Along the front, a cold air mass wedges underneath a warm air mass and lifts it. [Figure 3.7]. If the front is moving so that cold air is replacing warm air at the surface, it is a cold front. Warm air replacing cold air indicates a warm front. In both cases, the lighter warm air is lifted, usually more abruptly in a cold front.

Cold Fronts

Cold fronts usually enter Texas from the north or northwest. Less frequently they move from the west [Pacific fronts] or "back in" from the northeast. Weather associated with the fronts depends mainly upon the speed of the front and the moisture content and stability of the warm air mass which is being lifted. A rapidly moving cold front forces the warmer air mass to rise abruptly leading to the development of cumulus and cumulonimbus clouds and the possibility of thunderstorms. This type of cold front is characteristic of winter and spring. However, often in winter there is insufficient moisture to produce clouds and precipitation which results in temperature change without the occurrence of weather. Particularly severe cold outbreaks, often accompanied by dark clouds and occasional precipitation, are referred to as "blue northers." Some fronts pass through northwest Texas without incident but begin to produce clouds and weather when they encounter moist air in the central and eastern portions of the state. In either case, the cold front produces decreased temperatures. Coldest temperatures are experienced when continental polar air spreads southward out of Canada in association with an intense winter low pressure system. Less dramatic temperature falls are associated with Pacific cold fronts from the west or northwest or with fronts occurring during other seasons. The greatest probability of severe weather is in the warm season when warm, moist, unstable air is lifted along the frontal boundary. This situation is illustrated by the schematic cross section shown in Figure 4.2a. The warm air is lifted abruptly at the frontal boundary producing vertically developed clouds which may grow to severe proportions if the air mass is sufficiently unstable.

In late spring, early autumn and summer, cold fronts have difficulty penetrating the state. As a result, lifting along the front is not as abrupt and causes a shallow layer of cold air at the ground with warm air gliding over the top [Figure 4.2b]. This produces stratus-type clouds over a large area for an extended period of time, most frequently in the eastern half of the state.

The difference in weather patterns associated with the two cold-front types is primarily the areal extent of the clouds and precipitation. The steeper slope of the fast-moving front produces rapid upward motion at the front resulting in cumulonimbus clouds and occasional severe weather. The events in this case may be described as brief but violent, affecting a limited surface area. The slow-moving front, on the other hand, produces weather over a larger area for a longer period of time as precipitation from nimbostratus clouds falls through the shallow layer of cold air. If the front is weak or if the warm air is deficient in water vapor, clouds and rain may be absent. However, if the air gliding over the cold front is sufficiently unstable, it is possible for thunderstorms to develop as in the case of the faster-moving system. A typical sequence of events when a cold front passes includes an abrupt decrease in temperature and moisture content as winds shift from southerly to northerly and the development of clouds and weather depending on the moisture content and stability of the air. Whereas most wintertime cold fronts penetrate the entire state, the frequency of frontal passages decreases from north to south during other times of the year. This is particularly true in summer when cool fronts which advance through the High Plains seldom reach central Texas.

Occasionally in West Texas, weather is produced by the passage of a "dry line," a boundary separating dry air to the west from moist air to the east without an accompanying temperature change. Since dry air is heavier and

FIGURE 4.2a
A Fast-Moving Cold Front -
Unstable Air

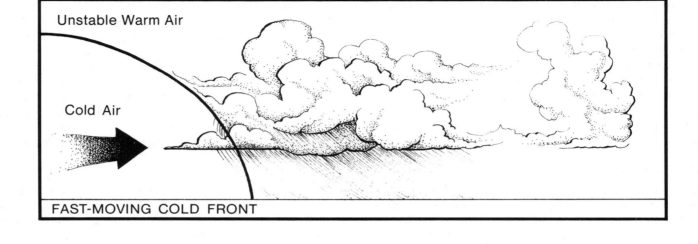

Unstable Warm Air

Cold Air

FAST-MOVING COLD FRONT

FIGURE 4.2b
A Slow-Moving Cold Front -
Stable Air

Stable Warm Air

Cold Air

SLOW-MOVING COLD FRONT

40

more dense than moist air, it forces the moist air upward in precisely the same way that cold air lifts warm air. This produces frontal-type weather without contrasting temperatures. Dry lines are characteristic of the warm season when warm, dry air is being drawn into West Texas from the arid southwest and moist air from the Gulf of Mexico dominates central and East Texas. Weather along the dry line is often severe.

Warm Fronts

A frontal boundary which moves such that warm air is replacing cold air at the ground is a warm front. Warm fronts enter Texas from the south and produce weather which is usually less abrupt than with cold fronts. Because the warm front slope is less steep the weather patterns tend to be similar to the patterns which develop with slow-moving cold fronts. As the warm air gradually moves upward over the cooler air at the ground, stratus clouds may develop which extend over a large area in advance of the front at the surface. In some instances frontal clouds may be observed as far as 1,000 miles ahead of the surface front.

A cross-sectional view of a warm front is shown in Figures 4.3a and 4.3b for stable and unstable conditions, respectively. The surface position of the front in both cases is located between Houston and Dallas. Weather is occurring well in advance of the front at the surface with clouds extending north into Oklahoma and beyond. As the front approaches, an observer would first see cirrus clouds followed by altostratus, stratus and nimbostratus.

When the front passes at the surface, winds will shift from northerly or northeasterly to easterly or southeasterly and the temperature and humidity will increase. Figure 2.13 illustrates the situation as seen on a weather map.

When the warm air which is being lifted is sufficiently unstable, thunderstorms may develop. This type of unstable warm front precipitation is common along the Texas coast during the cold season as warm, moist Gulf air streams up and over the cooler air at the ground. This regime may extend inland for hundreds of miles and may even affect the Texas Panhandle on rare occasions.

Most Texas warm fronts occur during winter and spring and are confined to southeast, east and central Texas. Warm fronts are infrequent during the warm season because the entire state is dominated by warm air and provides little opportunity for formation of a boundary between warm and cold air.

Stationary Fronts

During late spring and early summer, cold fronts which enter the High Plains often stall before they reach central Texas. These stalled fronts—called stationary fronts—tend to oscillate over a small area often producing large amounts of precipitation. Flooding often results as the rains persist for periods of several days. Stationary fronts are responsible for much of the late spring and early summer rainfall in the northern High Plains where many intruding cold fronts bog down. Frequently these fronts provide the lifting mechanism to generate severe weather.

FIGURE 4.3a
Warm-Front Weather in
Stable Air

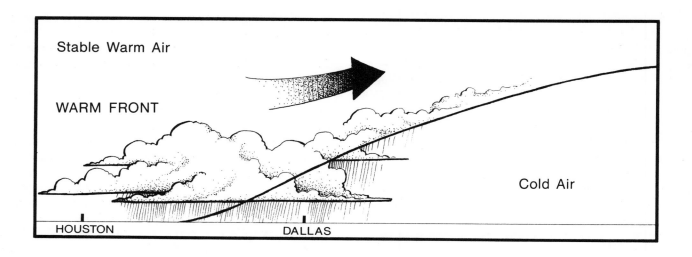

FIGURE 4.3b
Warm-Front Weather in
Unstable Air

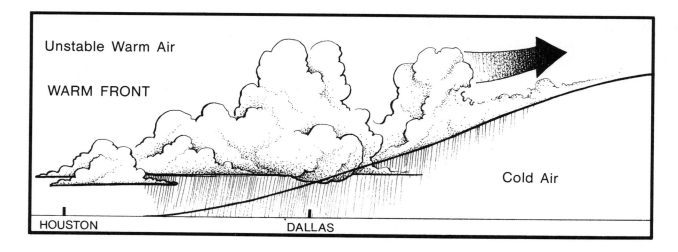

A thunderstorm is a local storm characterized by moderate to heavy rain, gusty and sometimes damaging surface winds, thunder and lightning. Heavy thunderstorms may become violent, reaching heights of 65,000 feet or more and producing hail and, occasionally, tornadoes. These are referred to as severe thunderstorms.

Thunderstorms originate in cumulus clouds, but only a few cumulus clouds develop into cumulonimbus, producing the intense upward and downward motion within the cloud characteristic of thunderstorms. The necessary ingredients for thunderstorm formation are abundant moisture, an unstable atmosphere (which usually means cold air at upper levels) and a triggering mechanism to initiate upward vertical motion. Several triggering mechanisms can be responsible for the development of Texas thunderstorms. Fronts can initiate upward motion as can low-level convergence or strong surface heating. Orographic effects in the mountains are responsible for showers and thunderstorms which produce a significant percentage of annual rainfall in the Trans-Pecos.

Texas thunderstorms are of two types: organized storms which form in continuous lines usually along fronts, and air mass thunderstorms which result from surface heating, convergence or orographic lifting within an air mass. Each storm is composed of one or more convection cells in various stages of development and decay. Each cell experiences a definite life cycle which may last from 20 minutes to more than an hour. The entire thunderstorm or cell cluster may last for several hours or more as new cells form and old ones dissipate. Typically there are three stages of development and decay for an individual cell: cumulus stage, mature stage and dissipating stage (Figure 4.4).

THUNDERSTORMS

FIGURE 4.4
Thunderstorm Stages of Development

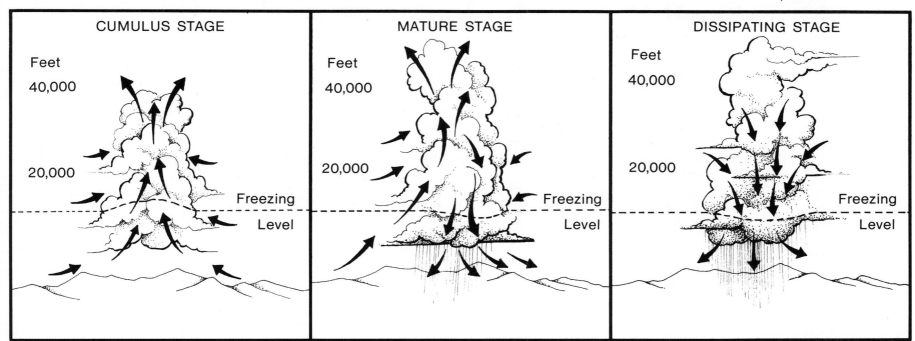

43

In the cumulus stage, the cloud is dominated by upward motion as it grows to become a cumulonimbus cloud. The end of the cumulus stage and beginning of the mature stage occurs when downdrafts are produced in association with precipitation falling out of the cloud. During the mature stage, both updrafts and downdrafts are occurring with heavy rain at the ground. Eventually, the downdraft expands and overcomes the updraft. This marks the beginning of the dissipating stage as the source of moisture and energy are cut off. This final stage is characterized by weak downdrafts and light rain as the storm gradually dies.

Air mass thunderstorms are most common during summer when strong surface heating causes rising columns of heated air to develop. These storms tend to grow during the middle or late afternoon when heating is most intense and may persist well into the night. In the High Plains they often become severe, producing heavy rain and hail. As noted previously, summer storms in the Trans-Pecos develop as a result of orographic lifting, primarily during the warm season. Air mass thunderstorms are also common along the Texas coast and tend to be highly localized, usually producing rainfall over limited areas.

Organized thunderstorms usually occur in association with fronts. These storms develop rapidly and tend to persist for long periods as new cells develop and grow. Heavy rain, damaging wind and hail often result. When the storms move slowly, abundant rainfall on a limited area may produce flooding both locally and downstream from the weather. Warm front thunderstorms may occur in east and central Texas. While infrequent, they are dangerous because they are usually embedded in a stratus overcast and are difficult to observe. Occasionally, when the air is sufficiently unstable, they may become severe.

Cold front thunderstorms are generally more severe and tend to occur in a line along the frontal boundary. During summer in the High Plains, spectacular thunderstorms are generated by slow-moving cold fronts which may become stationary. During spring and early summer, faster moving fronts produce a line of storms along the front and, frequently, a secondary line well in advance of the front called a squall line. These storms often become severe with heavy rain and hail, damaging winds and perhaps tornadoes.

Two by-products of thunderstorms are hail and lightning. Hail forms in thunderstorms when ice particles grow within the cloud to a size large enough to fall from the cloud and reach the ground before melting. Hail damage to crops, livestock and property results in millions of dollars in losses each year. The frequency of hail occurrence is highest in the High Plains, the low rolling plains, north central Texas and parts of the Edwards Plateau.

Lightning from thunderstorms produces damage in the millions of dollars and kills more than 200 persons each year on the average in the United States. It is caused by the mutual attraction of unlike electrical charges within a thunderstorm or between the storm and the ground. Most lightning occurs within a cloud or between one cloud and another. Cloud to ground lightning, which accounts for less than a third of all discharges, is a discharge between the negatively charged lower portion of a cloud and the positively charged ground. The sudden heating of the air along the path of the discharge produces expansion and compression waves resulting in thunder.

A tornado is a column of rapidly rotating winds in contact with the ground. It is nature's most violent storm with winds occasionally up to 325 miles per hour over an area typically less than a quarter of a mile in diameter. The largest tornadoes may reach a diameter approaching one mile whereas some may be less than 150 feet, particularly late in their life cycles.

According to an intensity scale [Table 4.1] developed by Dr. Theodore Fujita at the University of Chicago, tornadoes are classified in terms of observed damage. Approximate wind speeds are given for each category.

The pressure in a tornado funnel is substantially lower than the surrounding environment and since the tornado itself is relatively small, the pressure increases rapidly as one moves outward from the center of the core. This rapidly changing pressure [pressure gradient] over a short distance produces the strong winds which account for the storm's destructive nature. It is a common belief that the low pressure itself is responsible for structures exploding outward as a tornado passes over. However, it is the consensus of most structural engineers who have studied the problem that the damage is due to wind and only indirectly to pressure. The sequence of events begins when some part of the structure, usually a window or door, fails because of the high wind speeds. This allows wind to enter the structure producing an outward force on the roof and three walls. When this force is sufficiently strong, the building appears to explode as the roof and

TABLE 4.1
Fujita Tornado Intensity Scale

F0 [40-72 mph] Very weak tornado	Damage to branches, sign boards, TV antennae
F1 [73-112 mph] Weak tornado	Windows broken, roof shingles peeled, trailer homes overturned, trees snapped
F2 [113-157 mph] Strong tornado	Roofs removed, weak structures demolished, boxcars overturned, large trees uprooted
F3 [158-206 mph] Severe tornado	Walls removed from frame buildings, warehouses torn, cars lifted and rolled, trees leveled
F4 [207-260 mph] Devastating tornado	Frame houses leveled, trees debarked, structures lifted, missiles generated
F5 [261-318 mph] Incredible tornado	Steel-reinforced concrete structures damaged, automobile-sized structures transported over 100 yards
F6 [over 319 mph] Inconceivable tornado	Serious damage caused by large airborne debris, massive destruction

FIGURE 4.5
Wind Forces on a Building

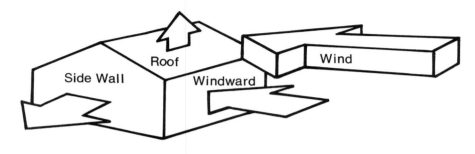

FIGURE 4.6
Conditions for Severe
Thunderstorms and Tornadoes

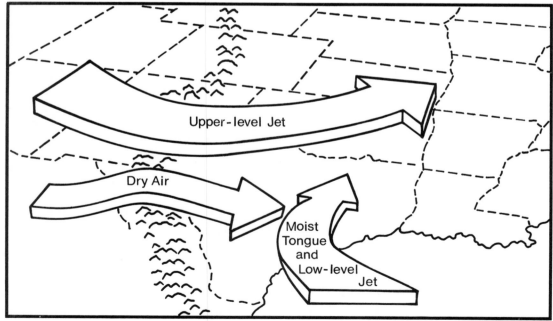

Figures 4.5 and 4.6 courtesy of Charles E. Merrill Publishing Company.

walls are forced outward [Figure 4.5]. Consequently, opening a window, as frequently instructed, when a tornado warning is issued will probably result in no more than a wet carpet. If the structure needs ventilation, the tornado will provide it.

Although most tornadoes are associated with severe thunderstorms, meteorologists do not agree on the precise mechanism for their development. Abundant moisture and extreme atmospheric instability are necessary. Other features prominently associated with tornadoes are a narrow core of high-velocity winds at upper levels—called a jet stream—and inflow of dry air from the west or southwest [Figure 4.6].

The tornado funnel usually develops from a distinct lowering of the cloud base known as a wall cloud. The wall cloud usually forms behind the area of heavy rain as shown in Figure 4.7. The storm in the diagram is moving from left to right. Warm moist air is being fed into the updraft causing explosive vertical cloud growth in this portion of the storm. When the cloud tops reach the extremely stable stratosphere, further growth is stopped and the cloud spreads out in an anvil. The wall cloud and tornado usually form below the maximum updraft in a region which is typically outside the area of heavy precipitation. Precipitation is associated with the downdraft and located in the forward quadrant of the storm. Cold gusty surface winds accompany the downdraft. If hail develops, it is usually on the tornado side of the precipitation area or between the rain and the tornado. The portion of the cloud base from which no rain or hail is falling is referred to as the rain-free base. New clouds are developing in this area in a pattern referred to as the flanking line. According to Figure 4.7, the sequence of events expected by an observer at the surface when a tornadic storm passes directly overhead would be: gusty winds associated with the downdraft, rain, hail and tornado. While this sequence would not hold true for every storm, it does represent an expected set of events.

Not all wall clouds produce tornadoes, but the presence of a wall cloud, particularly if it is rotating, signals a possi-

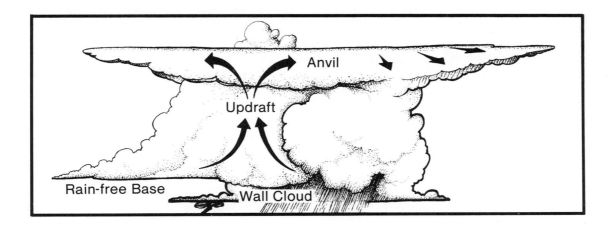

FIGURE 4.7
Schematic View of a
Tornado-Producing Severe
Thunderstorm

Anvil

Updraft

Rain-free Base

Wall Cloud

ble tornado. As air flows into the rotating wall cloud its speed increases significantly in much the same way that an ice skater quickens her rate of spin by drawing in her arms. Figure 4.8 shows the entire life cycle of a tornado which developed near Edmundson, Texas, near Plainview, on May 27, 1980. This entire sequence occurred during a period of less than 10 minutes. In Figure 4.8a, a wall cloud is visible and the first indication of tornado formation appears. The funnel is clearly visible in 4.8b and extends to the ground in 4.8e. In 4.8f, although the funnel cloud no longer extends to the ground, the wind circulation is evident from the debris below the funnel. The tornado circulation is still at the ground even though the funnel has lifted.

Texas, because of its location between the Gulf of Mexico and the Rocky Mountains, is a prime area for tornado formation. The pattern shown in Figure 4.6 is established by warm moist air flowing inland from the Gulf in an area adjacent to extremely dry air to the west. Initial lifting of the moist air is provided either by the dry line separating the air masses or by a cold front. Explosive growth is enhanced by atmospheric instability and the presence of an upper-level jet stream. This combination is responsible for the high frequency of tornadoes in the spring and early summer, particularly in the High Plains, low rolling plains,

and north central Texas. At this time of year, however, the entire state is susceptible. While most of the storms develop during the heat of the late afternoon and early evening, they can occur at any time. The time of minimum frequency extends from about midnight to noon. Most tornadoes are steered by strong southwesterly winds aloft and tend to move from southwest to northeast at speeds ranging from 15 to 40 miles per hour.

On the average, Texas experiences more than 125 tornadoes per year, more than half during the spring and early summer. A secondary maximum in tornado frequency occurs during late summer and early autumn in the southern third of the state. These tornadoes develop in conjunction with tropical storms and hurricanes which are active at this time of year. Tornadoes are extremely rare in southwest Texas and the Trans-Pecos Region. Table 4.2 lists some of Texas' most memorable tornadoes.

The National Weather Service issues a "tornado watch" when atmospheric conditions point to a potential tornado. The watch is issued for a specific area within which the tornado is most likely. A "tornado warning" means that a tornado has actually been sighted. When a warning is issued, persons in the vicinity should take immediate cover and those nearby should be on constant alert.

When seeking shelter from a tornado, common sense is a good guide. It's safer to be inside than outside, preferably in a basement. Otherwise, one should seek a small room such as a bathroom or closet in the central portion of the house. Outside walls should be avoided, as should windows and fireplaces. It is also a good idea to cover oneself with a mattress or other sturdy object to protect from falling objects and flying debris. If an underground tornado shelter is not readily accessible from the main residence or building, it is advisable to take shelter early. Many persons have been killed or injured while trying to reach shelter at the last minute. Since most deaths are caused by flying debris, it is best to stay where you are rather than risk a frantic dash to an out-door shelter. When out of doors, one should find a low spot on the terrain and lie flat. Automobiles and mobile homes should be avoided since they are easily tossed around by the tornadic winds.

Tornado damage cannot be escaped, but with adequate warnings and civil defense training, lives can be saved and injuries avoided.

FIGURE 4.8
Tornado Life Cycle

Photos courtesy of Carl Holland, Plainview, Texas.

FIGURE 4.8
continued

Date	Location	Deaths	Injuries	Property Damage ($)
Apr. 28, 1893	Cisco	23	93	400,000
May 15, 1896	in/near Sherman	76	?	225,000
May 18, 1902	Goliad	114	230	50,000
Apr. 26, 1906	Bellevue/Stoneburg	17	20	300,000
Mar. 23, 1909	Slidell	11	10	30,000
May 30, 1909	Zephyr	28	"many"	90,000
Apr. 9, 1919	Leonard/Ector/Rowena	20	45	125,000
Apr. 9, 1919	Henderson/Van Zandt/Wood Red River Counties	42	150	450,000
Apr. 13, 1921	Melissa	12	80	500,000
Apr. 15, 1921	Wood/Cass/Bowie Counties	10	50	85,000
May 4, 1922	Austin	12	50	500,000
May 14, 1923	Howard/Mitchell Counties	23	100	50,000
Apr. 12, 1927	Edwards/Real/Uvalde Counties	74	205	1,230,000
May 9, 1927	Garland	11	?	100,000
May 9, 1927	Nevada/Wolfe City/Tigertown	28	200	900,000
May 6, 1930	Bynum/Irene/Mertens/Ennis/Frost	41	?	2,100,000
May 6, 1930	Kenedy/Runge/Nordheim	36	34	127,000
Mar. 30, 1933	Angelina/Nacogdoches/ San Augustine Counties	10	56	200,000
June 10, 1938	Clyde	14	9	85,000
Apr. 28, 1946	Crowell	11	250	1,500,000
Jan. 4, 1946	Angelina/Nacogdoches	13	250	2,050,000
Jan. 4, 1946	Palestine	15	60	500,000
Apr. 9, 1947	White Deer/Glazier/Higgins	68	201	1,550,000
May 3, 1948	McKinney	3	43	2,000,000
May 15, 1949	Amarillo	6	83	5,310,000
Mar. 13, 1953	Jud/O'Brien/Knox City	17	25	600,000
May 11, 1953	Waco	114	597	41,150,000
May 2, 1957	Dallas (Oak Cliff)	10	200	4,000,000
May 15, 1957	Silverton	21	80	500,000
Apr. 3, 1964	Wichita Falls	7	111	15,000,000
June 2, 1965	Hale Center	4	76	8,000,000
Apr. 8, 1970	Clarendon	17	42	2,100,000
May 11, 1970	Lubbock	26	500	135,000,000
Mar. 10, 1973	Hill County	6	77	?
Apr. 10, 1979	Wichita Falls/Vernon	54	1807	442,000,000

From TEXAS WEATHER by George W. Bomar, Copyright© 1983 by the University of Texas Press. Used by permission of the publisher.

Hurricanes, the most destructive of all storms, are similar to tornadoes in that their primary characteristics are low pressure and high wind speed. But the two storms differ markedly in size and duration. A typical hurricane is 375 miles or more in diameter and lasts for several days. However, some can be much larger and last for a week or more. The pressure at the center of a hurricane is extremely low and increases rapidly with distance from the center. This strong pressure gradient produces wind speeds which occasionally exceed 175 miles per hour. At the center of the hurricane is the "eye," a usually symmetrical area 10 to 15 miles in diameter. In some large storms the eye may have a diameter of 20 miles or more. Within the eye, skies are clear or partly cloudy and winds are calm or very light. Vertical motion in the eye is downward which explains the stable conditions and lack of clouds. Strong upward motion occurs outside the eye producing thunderstorms and heavy rain. Heaviest rainfall is in the right front quadrant of the storm which is also the region in which tornadoes are most likely to form. A cross section of a hurricane illustrating the pattern of vertical motion is shown in Figure 4.9.

Whereas the precise mechanism responsible for hurricane development is not clearly defined, the storms develop over tropical oceans between 5° and 20° from the equator and owe their existence to energy released when the water vapor provided by the ocean is lifted in the storm and condensed. Thus, the storms are fueled by the input of moist air at low levels. Of equal significance are the upper-level winds responsible for transporting this upward moving air away from the storm, producing a heat engine which allows the input of moist air to continue. Rotation of the storm is due to the rotation of the earth (Coriolis force) resulting in counterclockwise circulation in the Northern Hemisphere. Hurricanes do not form within 5° latitude of the equator because the Coriolis force is not sufficiently strong at this low latitude to initiate rotation.

Hurricanes develop in stages. The first stage, called a "tropical disturbance," is characterized by a disorganized region of low pressure and some cloud development. This becomes a "tropical depression" when a symmetrical low pressure center develops. Further lowering of the pressure, accompanied by an increase in wind speed to at least 39 miles per hour, produces a "tropical storm." The tropical storm becomes a hurricane when its wind speed equals or exceeds 74 miles per hour.

North Atlantic or Caribbean hurricanes which affect Texas develop in the easterly trade winds and move in a westerly direction at an average speed of about 15 miles per hour. As the storms approach the continent they curve northward and become embedded in the prevailing westerlies. Accompanying this change in direction from westerly to easterly is an increase in speed to as high as 60 miles per hour. When the hurricane moves ashore, its energy source is lost. This loss of energy input coupled with the increased frictional effect of the land surface causes the storm to lose intensity and die. In some cases the hurricane may encounter a mid-latitude front as it moves ashore and become a heavy rainmaker producing flash floods over a wide area. Other storms may be relatively dry. In many storms the rapid dissipation that occurs with landfall, the time when the leading edge of the storm moves ashore, is accompanied by an outbreak of tornadoes.

HURRICANES

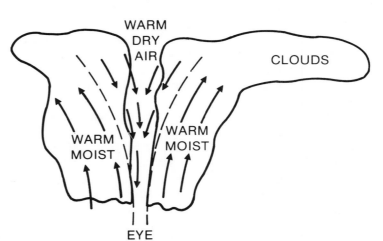

FIGURE 4.9
Vertical Motion in a Hurricane

Although the hurricane season in the Gulf of Mexico extends from June through October, storms rarely develop during early June and late October. Early and late season hurricanes usually have their origins in the Caribbean or Gulf of Mexico. Storms which develop during July, August and early September are most frequently initiated in the tropical waters of the North Atlantic ocean and drift westward into the Caribbean and Gulf.

From 1871 to 1982, 42 hurricanes have made landfall in Texas and 14 others have caused damage in the state. During that same period, 30 tropical storms entered the state and 13 others came close enough to produce some effects. The interannual variability is very large. Whereas some years have provided two or more tropical storms or hurricanes, other years have produced none. Table 4.3 lists the hurricanes which have hit the Texas coast during the period 1941 to 1982. Note that the longest period without a hurricane entering the state ended on August 10, 1980, when Hurricane Allen moved ashore near Port Mansfield and tracked across South Texas (Figure 4.10). The storm was generated over the warm tropical Atlantic and moved westward through the Caribbean killing almost 300 persons and inflicting massive property damage. It then struck the Yucatan Peninsula with 172-mile-per-hour winds before entering the Gulf of Mexico, reintensifying and heading for the Texas coast. Allen was one of the most intensive storms ever in the Gulf with a low pressure of 899 mb. Although it lost intensity before moving ashore, it buffeted the South Texas coast with 110-mile-per-hour winds and produced several damaging tornadoes. Coastal areas were under water as a result of the storm surge (tide) and a large area of South Texas was flooded by heavy rainfall.

Figure 4.11 shows the tracks of significant hurricanes which have struck the Texas coast since the turn of the century. Of course many other storms have affected the state during this period but are not illustrated because they did not make landfall.

Each hurricane is unique. The rainfall is usually showery with large variations within a particular storm and from storm to storm. Generally, most of the rain falls near the region of maximum winds, with squalls out 20 to 200 miles from the eye.

One of the wettest hurricanes to enter Texas was Beulah, September 19-23, 1967. Most of the region from Matagorda Bay northwest to San Antonio and south to Laredo received at least 10 inches of rainfall; amounts up to 30 inches were recorded in some locations. Celia, which affected Texas during the period August 2-5, 1970, was a relatively dry storm which turned out to be one of the costliest hurricanes in Texas' history. The worst natural disaster in U.S. history occurred at Galveston, Texas, when a hurricane moved ashore on September 8, 1900, catching the city completely unprepared and killing 6,000 people.

The destructiveness of hurricanes is attributable to three things: wind; storm surge; flooding from heavy rains. The primary cause of damage in most storms is the storm surge, a wall of water produced by winds as the storm approaches the coast. This tide of water, which can reach a height of 12 feet or more, sweeps ashore producing damage due to both the force of the water's impact and flooding. If torrential rains continue, flooding is also a threat all along the path of the storm as it moves inland. The shallow continental shelf region off the Texas coast enhances the probability of a damaging storm surge as a storm moves ashore. In addition, the water collection and runoff systems in many cities are not capable of handling the large amount of precipitation which accompanies hurricanes and tropical storms. The upper coast of Texas, particularly in the Houston-Galveston area, is an example of a region easily devastated by heavy rains. The flat topography coupled with subsidence (lowering of the height above sea level) makes this region particularly susceptible to flooding and water damage.

Wind damage is the other factor to be considered. In the strongest storms, wind force can cause total destruction of some structures and serious damage to most. High wind speeds are sustained for a longer period of time than in tornadoes but are not usually as great in magnitude. Often, the generation of tornadoes within the hurricane produces even more disastrous winds and consequent damage.

Although our ability to forecast hurricane development has not increased appreciably during the past twenty years or so, our ability to track a storm has. With satellites, radar, reconnaissance aircraft and ship reports, the storm's location, intensity and path can be carefully monitored and reported. In the case of Hurricane Allen [Figure 4.10], advance warning allowed 200,000 persons to be evacuated from the coastal areas.

The Hurricane Warning Office in New Orleans is responsible for monitoring tropical storms and hurricanes and issuing watches and warnings. A hurricane watch is the first alert issued when a storm threatens a coastal area. The hurricane watch statement usually includes small craft advisories as well. A hurricane warning is issued when a storm is expected to strike an area within a 24-hour period. Specifically, the warning indicates that within 24 hours the area is expected to encounter sustained winds of 74 miles per hour or higher and/or high water and exceptionally high waves.

TABLE 4.3
Statistics on Major Hurricanes That Hit the Texas Coast, 1941-82

Year	Name	Date*	Time [CDT]*	Location*	Maximum Winds [mph]*	Lowest Pressure [mb]	Damage [millions of $]	Deaths
1941		Sep 23	evening	Matagorda			$ 6.5	4
1942		Aug 21	dawn	Galveston				
		Aug 30	early A.M.	Matagorda Bay	115		26.5	8
1943		Jul 27	late A.M.	near Galveston	104		16.5	19
1945		Aug 27	about noon	Aransas Bay	135		20.1	3
1947		Aug 24	early P.M.	Galveston				
1949		Oct 3	early A.M.	Freeport	135		6.5	2
1957	AUDREY	Jun 27	9 A.M.	near Sabine Pass	100	958	8.0	9
1959	DEBRA	Jul 25	6 A.M.	Dickinson		986	7.0	0
1961	CARLA	Sep 11	2 P.M.	Port O'Connor	175	935	408.3	34
1963	CINDY	Sep 17	10 A.M.	High Island	50	996	11.7	3
1967	BEULAH	Sep 20	6 P.M.	near mouth of Rio Grande	115	951	200.0	15
1970	CELIA	Aug 3	3:30 P.M.	Corpus Christi Bay	130	945	453.7	11
1971	FERN	Sep 10	11 A.M.	Rockport	91	979	30.2	2
1980	ALLEN	Aug 10	1 A.M.	Port Mansfield	115	945	600.0	2

* at the time of landfall
\# no names assigned prior to 1953

From TEXAS WEATHER by George W. Bomar, Copyright© 1983 by the University of Texas Press. Used by permission of the publisher.

Courtesy of National Environmental Satellite Service.

FIGURE 4.10
Hurricane Allen

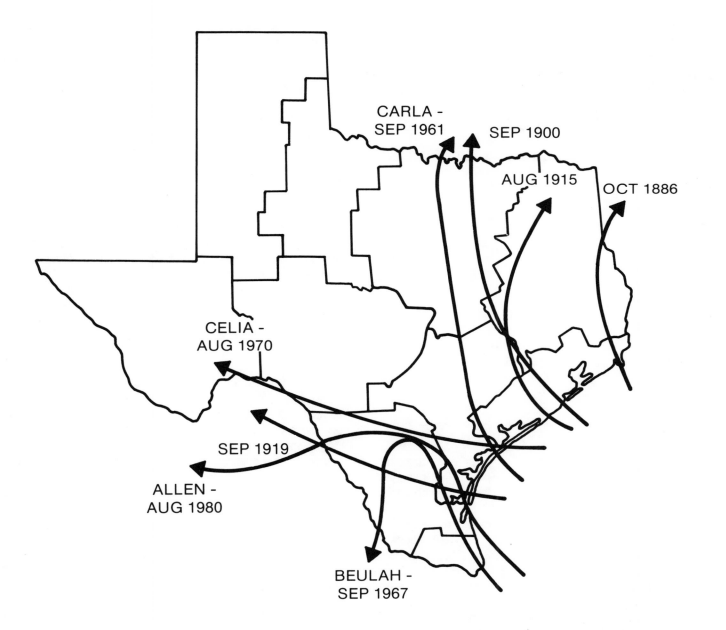

CARLA -
SEP 1961

SEP 1900

AUG 1915

OCT 1886

CELIA -
AUG 1970

SEP 1919

ALLEN -
AUG 1980

BEULAH -
SEP 1967

FIGURE 4.11
Tracks of Significant Hurricanes
Which Have Struck the Texas
Coast

DUST STORMS

One of the dustiest locations on the globe is the South Plains of Texas. Considering all stations with meteorological records, no place in North America observes more hours of blowing dust than Lubbock, Texas. Table 4.4 shows the average number of days and hours of blowing dust for each month of the year at Lubbock.

TABLE 4.4
Blowing Dust at Lubbock, Texas

Month	Days	Hours
January	3.4	12.3
February	3.7	17.8
March	6.2	36.8
April	5.9	30.0
May	4.1	15.1
June	2.8	7.9
July	0.8	1.1
August	0.6	0.7
September	0.5	1.1
October	0.8	1.9
November	1.9	6.9
December	3.2	15.2
Annual Total	33.8	146.8

While table 4.4 is informative, it is not complete because of the large year-to-year variation. For instance, the number of hours of blowing dust during March varied from zero in 1958 to 121 hours in 1954. During the same period, the total number of hours of annual blowing dust varied from 44 hours in 1963 to 438 hours in 1953. Figure 4.12 shows the frequency of occurrence of blowing-dust hours for the period 1947-1981. The graph indicates that during the 35-year period, three years recorded less than 50 hours of blowing dust whereas 10 years had between 50 and 100 hours of blowing dust. The only year which experienced more than 400 hours of blowing dust was 1953. Approximately one-half of the years

recorded more than 145 hours of blowing dust. During that period, the maximum occurrence of blowing dust was in either March or April in 25 of the years, in December in four years and in January, February or May in each of two other years. The fewest hours of blowing dust occurred in July, August and September.

The factors required to produce a dust storm are wind and dust which is free to blow. In the Texas South Plains, agricultural practices create a dust source. The major crop in the area is cotton which is grown in summer and harvested in fall and early winter. During late winter and spring the fields are barren. In addition, several months of dry weather normally precede the late winter-spring season which further increases the chance that the dust will blow. When unusually wet winters occur, blowing dust can be reduced appreciably.

Dust devils—small whirlwinds which develop on hot dry days—can be responsible for stirring up small amounts of local dust. Winds in these storms can range from 20 miles per hour to perhaps 100 miles per hour on rare occasions, but the size and intensity of a dust devil are not sufficient to produce a major dust storm.

Dust storms on the Texas South Plains are usually related to one of four causes. In the order of increasing severity they are thunderstorm outflow, fronts, low pressure troughs and deep low pressure systems.

During the summer months dust is often picked up briefly by outflowing air from thunderstorms. Although visibility may be reduced to nearly zero, the dust is usually short lived, lasting only a few minutes to half an hour. Cold fronts can often produce dust episodes lasting several hours. Most of the dust in these cases is produced by brisk southwesterly winds in advance of the front. If the northerly winds behind the front have sufficient speed, the dust may continue to blow for an hour or so after the front has passed. Low pressure troughs, both at the surface and at upper levels, produce substantial portions of South Plains dust. In this case, the rapid heating of the day often produces sufficient mixing to bring high-speed winds at upper levels down to the surface. A day which begins with a clear, calm morning can deteriorate to a windy, dusty afternoon. The most severe type of dust storm results from deep surface low pressure systems which frequently form on the leeside of the Colorado Rockies. These systems often produce westerly winds exceeding 50 miles per hour for extended periods of time, up to 36 hours.

The impact of blowing dust on the Texas South Plains is significant. Storms can cause considerable damage to crops and agricultural land not to mention extreme discomfort to residents. In the most severe storms, the loss of top soil alone can reach millions of dollars. Although, according to the National Weather Service, a dust storm is one in which dust is raised to moderate heights and reduces visibility to 5/8ths of a mile or less, South Plains residents can attest to the fact that in some storms it is indeed difficult to see your hand in front of your face.

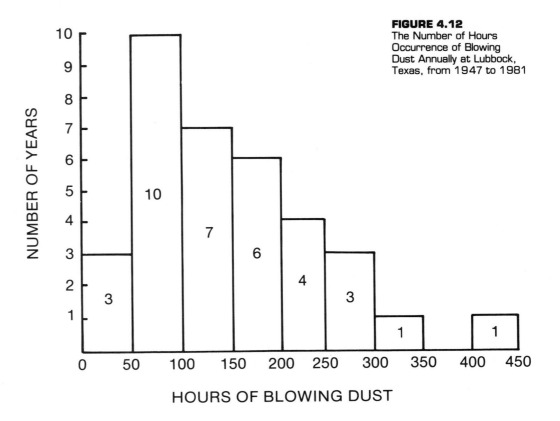

FIGURE 4.12
The Number of Hours Occurrence of Blowing Dust Annually at Lubbock, Texas, from 1947 to 1981

Chapter 5 / TEXAS CLIMATE

Climate is the summation of weather events that occur at a particular place over a long period of time. It represents the net effect of the weather patterns previously discussed. A common misconception is to define a region's climate as merely the average state of the atmosphere. Not only are average temperature, precipitation, wind and humidity important, but also daily, seasonal and yearly variations. Extremities of temperature and precipitation are particularly important when determining what crops can be grown, how buildings must be designed and the manner in which general human activities can be conducted.

An excellent analogy to the atmosphere and the importance in considering variation of temperature and precipitation when defining a climate is the story of the man with one foot in a bucket of ice and the other on hot coals. On the average, he is comfortable. That Lubbock receives an average precipitation of about 17¾ inches per year takes on a whole new meaning when one considers the individual years from which the average was computed. Since 1912 there have been ten years in which the total precipitation was less than 12 inches and ten years when precipitation was greater than 25 inches. In 1941, Lubbock received 40.55 inches of precipitation.

This chapter considers the Texas climate with particular emphasis on temperature and precipitation, the two most important climatic elements. Both averages and extremes are considered for the state as a whole, its ten climatic regions and selected stations throughout the state. Averages for the various climatic regions are for the period 1951-1980 and are based on the following stations for each region:

High Plains	Amarillo, Big Spring, Dalhart, Lubbock, Midland-Odessa, Seminole, Spearman, Hereford.
Low Rolling Plains	Abilene, Childress, Haskell, Matador, Seymour, Snyder, Vernon, Wichita Falls.
North Central	Brownwood, Dallas-Fort Worth, Gainesville, Graham, Greenville, Mineral Wells, Taylor, Waco.
Eastern	Clarksville, College Station, Huntsville, Lufkin, Marshall, Mount Pleasant, Palestine, Sulphur Springs.
Trans-Pecos	Alpine, Chisos Basin, El Paso, Fort Stockton, Mount Locke, Pecos, Presidio, Wink.
Edwards Plateau	Brady, Del Rio, Fredericksburg, Lampasas, McCamey, San Angelo, Sonora, Uvalde.
South Central	Austin, Beeville, Brenham, Corpus Christi, Flatonia, Hallettsville, San Marcos, San Antonio.
Upper Coast	Angleton, Galveston, Houston, Liberty, Matagorda, New Gulf, Port Arthur, Victoria.
Southern	Alice, Cotulla, Eagle Pass, Encinal, Falfurrias, Laredo, Poteet, Rio Grande City.
Lower Valley	Brownsville, Harlingen, McAllen, Raymondville.

Locations of the above stations are illustrated in Appendix 2. Appendix 5 provides average monthly and annual temperature for each station. Precipitation data for each station are given in Appendix 6.

TEMPERATURE

Figure 5.1 shows the distribution of average annual temperatures (1951-1980) for the state. As expected, temperatures decrease with latitude from south to north. The average temperature in far southwest Texas exceeds 74 °F while that in far northwest Texas is less than 54 °F. The temperature change across the state is fairly regular except in the mountains of far West Texas where higher elevation causes lower temperatures than in other areas of the state at the same latitude.

Average temperatures for the cold and warm seasons are illustrated by maps for January and July in Figures 5.2 and 5.3. With the exception of summer (Figure 5.3), the pattern in each season looks much like the annual pattern with highest temperatures in South Texas and the lower Valley decreasing regularly from south to north to minimum temperatures in the High Plains. This pattern is denoted by the east-west orientation of the isotherms (lines of constant temperature). The isotherm pattern is noticeably different in July (Figure 5.3). In this case, a distinct tongue of warm air (shaded area in Figure 5.3) extends either side of a line from Laredo to Austin to Dallas-Fort Worth to Wichita Falls causing the isotherms to have a north-south orientation. This band of high temperatures is characteristic of Texas summers with average temperatures higher than 84 °F and a maximum near 88 °F along the southwest border north of Laredo.

Table 5.1 shows the seasonal variation of average temperatures among the state's ten climatic regions. The warm tongue of summer air shown by the shaded area in Figure 5.3 is evident in the table and extends from the southern region (86 °F) through the south central (84 °F) and north central (85 °F) regions and west to portions of the Edwards Plateau (84 °F) and the Low Rolling Plains (84 °F). The warmest summer temperatures are in the southern region while the coldest winter temperatures are in the High Plains. This pattern is consistent, particularly in winter. Texas' coldest average temperature has been recorded in the High Plains for each of the last 20 years. The warmest average temperature has been recorded in the southern region for 16 of the last 20 years.

In order to more accurately record temperature distribution, it is necessary to study both the average temperature range and the extreme values which have occurred. The daily range of temperature is the difference between the maximum and minimum temperatures on

TABLE 5.1
Mean Monthly and Annual Temperature (°F)

Climatic Region	January	April	July	October	Annual
High Plains	38	59	80	61	59
Low Rolling Plains	41	64	84	64	63
North Central	44	65	85	67	65
Eastern	45	65	83	67	65
Trans-Pecos	46	65	80	64	64
Edwards Plateau	47	68	84	68	66
South Central	51	70	84	71	69
Upper Coast	53	69	83	71	69
Southern	54	74	86	73	72
Lower Valley	59	75	84	75	73

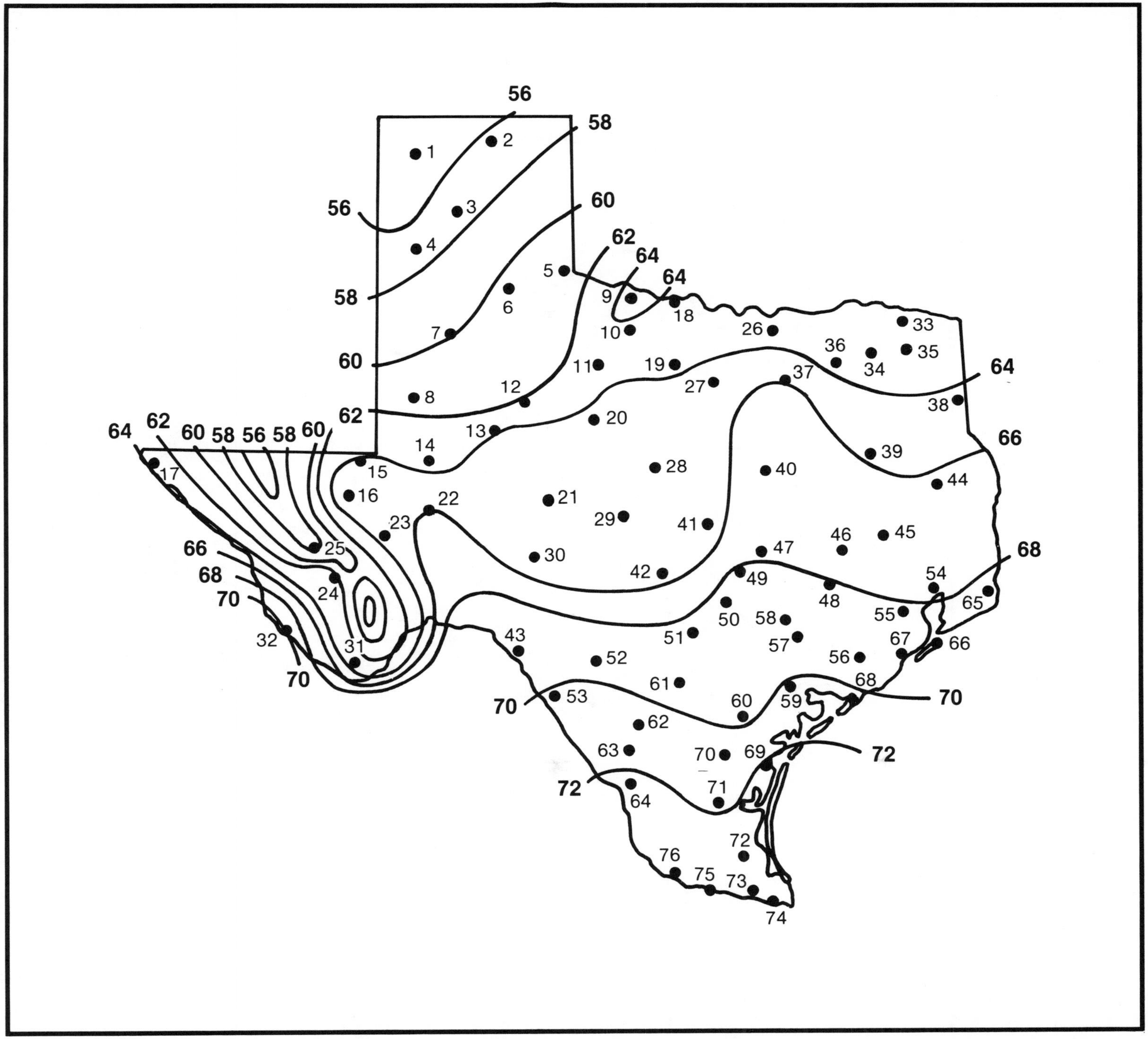

FIGURE 5.1
The Distribution of Mean
Annual Temperature (Deg F)

FIGURE 5.2
The Distribution of Mean
January Temperature
(Deg F)

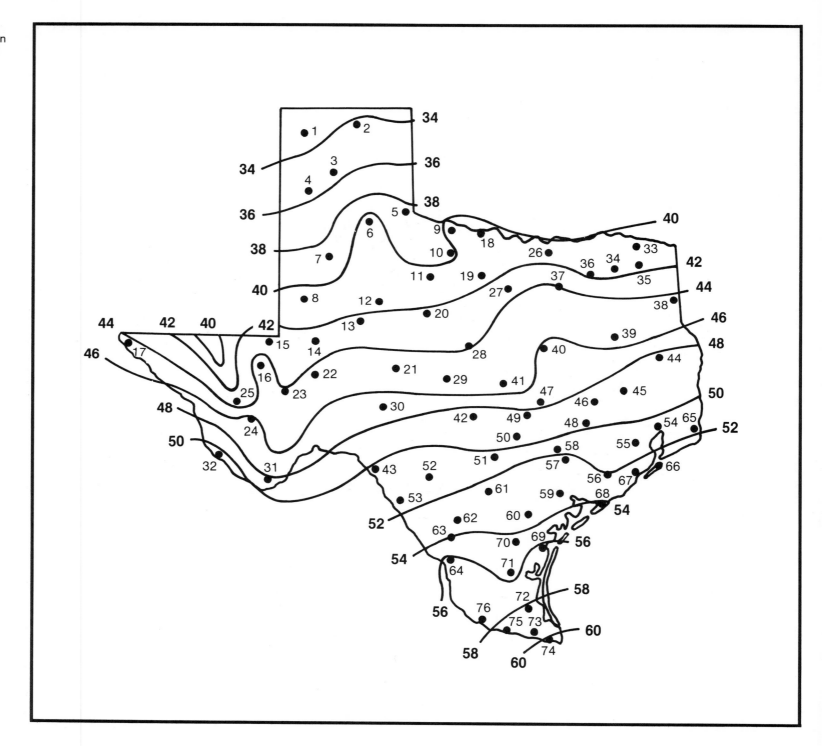

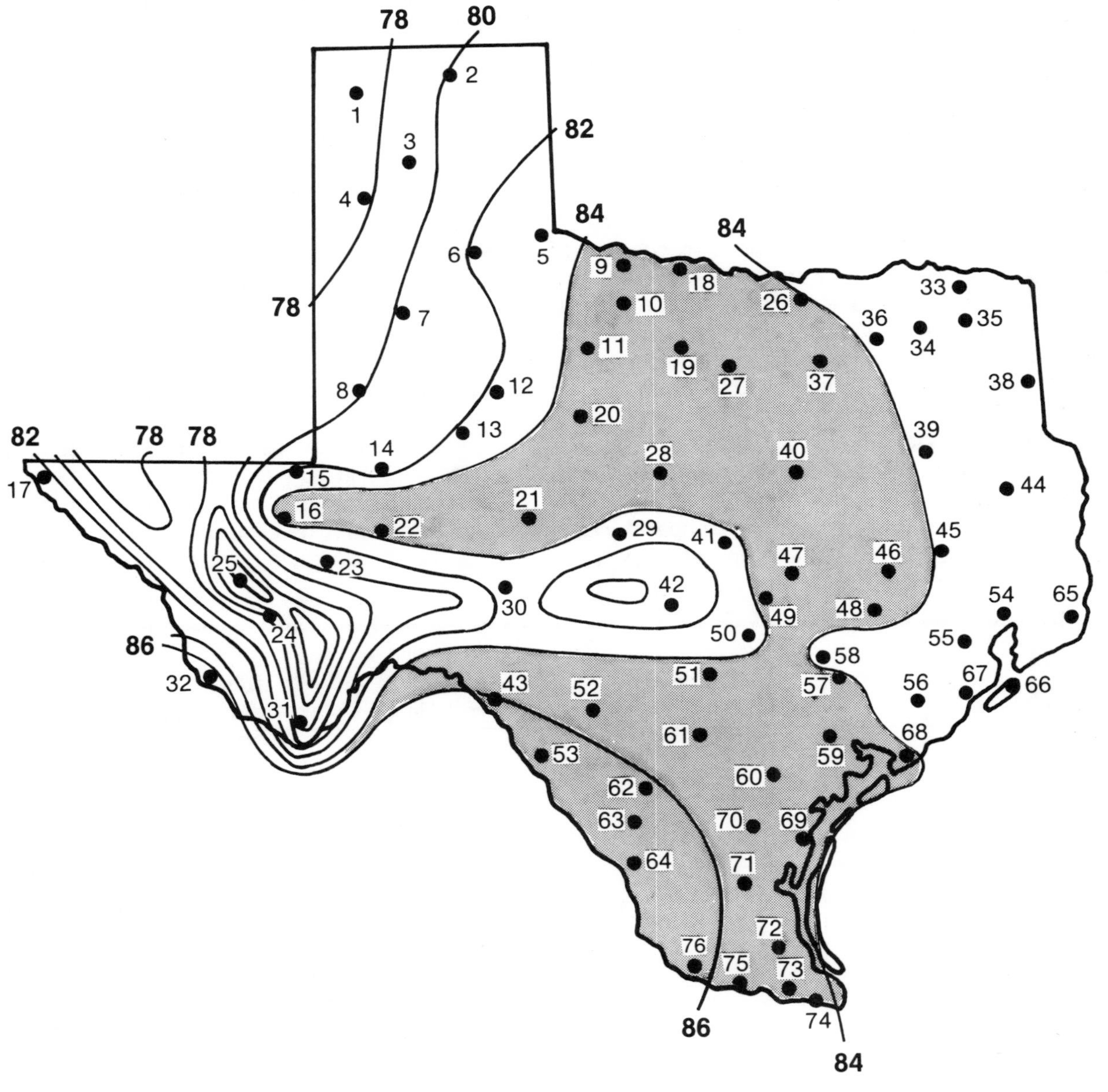

FIGURE 5.3
The Distribution of Mean
July Temperature (Deg F)

that day. Daily ranges may be averaged to produce an average monthly temperature range. Table 5.2 provides average temperature ranges for January, April, July and October for each climatic region. Average maximum and minimum temperatures are included for comparison.

A large temperature range is indicative of a continental climate, which is relatively dry and not influenced by large bodies of water. Examples are the High Plains and Trans-Pecos where low humidities allow maximum input of radiation during the day and rapid cooling at night because of less absorption of radiation by water vapor. A small temperature range indicates a humid maritime climate and usually reflects the influence of nearby water, in this case the Gulf of Mexico. Maritime climates dominate the upper coast, lower Valley and south central regions. As an example of the effect of humidity, compare the High Plains and the upper coast in summer. Maximum temperatures during the day differ by only one degree. However, due to high humidity along the coast, the temperature falls to only 74 °F at night, while overnight minimum temperatures in the High Plains are 66 °F. Thus, the temperature range is considerably higher in the High Plains due to its semi-arid climate.

Extreme temperatures, though infrequent, can have a profound effect on human activities and are an important factor in a region's climate. Table 5.3 gives the highest and lowest temperatures which have been recorded at a large number of stations throughout the state. Temperatures above 100 °F are observed in all regions of the state. Temperatures averaged over the entire warm season are highest in the south even though single-day maximum temperatures are usually highest in the Trans-Pecos. Summer averages are not as high in the Trans-Pecos because of the significant overnight cooling of the dry atmosphere. Coldest temperatures, either on the basis of seasonal averages or daily minimums, occur invariably in the High Plains.

Maximum and minimum temperatures for the state are given in Table 5.4 for each month. Since many of the records were established in remote areas whose names are unfamiliar, the climatic region in which the station is located is included in the table.

TABLE 5.2
Temperature Range [°F]

Climatic Region	January Max	January Min	January Range	April Max	April Min	April Range	July Max	July Min	July Range	October Max	October Min	October Range
High Plains	53	24	29	74	44	30	93	66	27	75	46	29
Low Rolling Plains	54	28	26	77	50	27	97	71	26	78	51	27
North Central	55	32	23	77	53	24	97	73	24	79	54	25
Eastern	56	34	22	77	54	23	94	71	23	80	54	26
Trans-Pecos	60	31	29	81	49	32	93	67	26	79	50	29
Edwards Plateau	61	33	28	81	54	27	96	71	25	81	54	27
South Central	62	40	22	80	59	21	95	74	21	82	59	23
Upper Coast	62	43	19	78	60	18	92	74	18	82	60	22
Southern	67	42	25	86	61	25	98	74	24	85	60	25
Lower Valley	70	48	22	85	65	20	95	74	21	86	64	22

Heat waves and cold waves which persist for long periods can be particularly devastating. During the severe winter of 1976-77, the northern High Plains experienced more than 50 consecutive freeze days. Much of the Trans-Pecos experienced 75 or more freeze days, although not consecutively. At the other extreme, Dallas-Fort Worth recorded 69 days with temperatures above 100 °F during the summer of 1980. At Wichita Falls temperatures rose above 100 °F on 79 days during this same heat wave.

TABLE 5.3
Temperature Extremes for Selected Stations [°F]

Station	Max	Min	Station	Max	Min
Abilene	111	−9	Houston	107	5
Alpine	106	−2	Laredo	115	16
Amarillo	108	−16	Lubbock	109	−17
Austin	109	−2	Lufkin	110	−2
Big Spring	110	−7	McAllen	107	13
Brownsville	104	12	Midland-Odessa	109	−11
Brownwood	111	−2	Mineral Wells	114	3
Childress	115	−7	Pecos	118	−9
Chisos Basin	103	−3	Port Arthur	107	11
College Station	109	−3	San Angelo	111	1
Corpus Christi	105	11	San Antonio	107	0
Dalhart	110	−21	Sonora	109	−8
Dallas-Fort Worth	113	−8	Uvalde	114	6
Del Rio	111	11	Victoria	110	9
El Paso	112	−8	Waco	112	−5
Galveston	101	8	Wichita Falls	117	−12

TABLE 5.4
State Maximum and Minimum
Temperature Records

Month		T(°F)	Station	Climate Region	Date
January	Max	98	Fort McIntosh	Southern	18 Jan 1914
		98	Laredo	Southern	17 Jan 1936
	Min	−22	Spearman	High Plains	4 Jan 1959
February	Max	104	Fort Ringgold	Southern	26 Feb 1902
	Min	−23	Tulia	High Plains	12 Feb 1899
		−23	Seminole	High Plains	8 Feb 1933
March	Max	108	Fort Ringgold	Southern	14 Mar 1902
		108	Rio Grande City	Southern	30 Mar 1954
	Min	−12	Romero	High Plains	1 Mar 1922
		−12	Spearman	High Plains	6 Mar 1948
April	Max	113	Falcon Dam	Southern	10 Apr 1963
	Min	5	Romero	High Plains	2,6 Apr 1936
May	Max	115	Encinal	Southern	24 May 1925
		115	Fort McIntosh	Southern	7 May 1927
		115	Boquillas Ranger Station	Trans-Pecos	8 May 1952
	Min	15	Tulia	High Plains	1 May 1909
June	Max	118	Pecos	Trans-Pecos	29 Jun 1968
	Min	32	Tulia	High Plains	3 Jun 1917
July	Max	119	Tilden	Southern	2 Jul 1910
	Min	40	Claytonville	Low Rolling Plains	4,9 Jul 1970
		40	Mount Locke	Trans-Pecos	13 Jul 1970
		40	Mount Locke	Trans-Pecos	16 Jul 1972
August	Max	120	Seymour	Low Rolling Plains	12 Aug 1936
	Min	39	Plemons	High Plains	26 Aug 1910
September	Max	115	Boquillas Ranger Station	Trans-Pecos	1 Sep 1952
	Min	29	Mount Locke	Trans-Pecos	29 Sep 1945
October	Max	109	Victoria	Upper Coast	11 Oct 1926
	Min	8	Fort Hancock	Trans-Pecos	28 Oct 1970
November	Max	101	Fort McIntosh	Southern	17 Nov 1906
	Min	−10	Stratford	High Plains	28 Nov 1976
December	Max	98	Encinal	Southern	1,2 Dec 1921
		98	Cotulla	Southern	6 Dec 1951
		98	Carrizo Springs	Southern	7 Dec 1951
	Min	−16	Booker	High Plains	11 Dec 1932

Figure 5.4 shows the distribution of average annual precipitation. Precipitation decreases from a maximum of more than 55 inches in far southeast Texas and along the upper coast to a minimum of less than 10 inches in the Trans-Pecos region of far West Texas. The decrease from east to west is fairly regular and most pronounced in the eastern half of the state. The only irregularities are the bulge in the 30-inch isohyet [a line connecting points with equal amounts of precipitation] along the Balcones Escarpment and the heavier precipitation in the mountains of West Texas as a result of orographic effects.

Maps representing winter [January], spring [May], summer [July] and fall [September] precipitation patterns are shown in Figures 5.5, 5.6, 5.7 and 5.8, respectively. The months of May and September, rather than April and October, are used to illustrate spring and fall because they are more representative of the heavy precipitation which develops during these seasons. In January and May the regular decrease in precipitation from east to west is similar to the annual pattern. January precipitation decreases from more than four inches in the east to less than an inch over the entire western half of the state. Precipitation in May decreases from more than five inches in the east to less than an inch in the Trans-Pecos region. The pattern in July represents a significant change. The maximum precipitation of greater than five inches occurs in southeast Texas with rainfall evenly distributed over the remainder of the state. This even distribution is not unexpected since most of the summer rainfall results from scattered, isolated storms which are set off by strong surface heating. In the mountains of West Texas, topographic effects aid storm growth and development. The pattern becomes more regular in September with heavy tropical rains observed along the coast and less than two inches of precipitation in far West and northwest Texas.

Table 5.5 shows average monthly and annual precipitation among the ten climatic regions. The averages have been computed using the same group of stations for each climatic region which were used in the previous section for

temperature. The upper coast, with 45.54 inches, receives the highest average precipitation followed closely by the eastern region [43.33 inches]. Smallest average amounts are in the Trans-Pecos [12.53 inches] and the High Plains [16.96 inches]. Since values in the table are averages over several stations in each region, extreme values for individual stations will be discussed separately. Appendix 6 provides average precipitation for selected cities.

Table 5.5 is summarized by Figure 5.9 which shows the annual distribution of average monthly precipitation for each climatic region. It is important to recognize that the causes of precipitation are different at different times of the year. Over most of Texas the heaviest rainfall occurs in the spring and fall as shown in Figure 5.9. The heavy rainfall in April and May is usually the result of frontal lifting of warm moist air. During this time of year when water vapor is plentiful in the atmosphere and the necessary vertical motion is provided, precipitation follows almost without exception.

Summer rainfall is usually in the form of scattered showers or thunderstorms which depend mainly on daytime heating, low-level moisture, and the absence of high pressure and downward vertical motion. In the Trans-Pecos, summer showers are enhanced by topographical effects. General rainfall covering large areas and where duration is greater than a few hours is associated, even during the summer season, with fronts which occasionally move into Texas, particularly into the northern High Plains. During the summer, however, high pressure dominates the state and inhibits the development of precipitation in most areas.

Heavy late summer and early autumn rainfall over most of Texas is mainly the result of tropical disturbances moving northward and westward from the Gulf of Mexico. This is particularly noticeable by the September precipitation maximum along the upper coast and in the lower Valley, southern, and south central climatic regions. Late autumn and winter rains are usually the result of warm moist Gulf air overrunning continental polar air which is associated

TABLE 5.5
Average Precipitation (inches)

Climatic Region	Jan	Feb	Mar	Apr	May	Jun	Jul	Aug	Sep	Oct	Nov	Dec	Ann
High Plains	0.45	0.57	0.79	1.09	2.49	2.38	2.36	2.28	1.95	1.49	0.66	0.45	16.96
Low Rolling Plains	0.76	0.89	1.22	2.11	3.68	2.69	2.17	2.29	3.01	2.41	1.12	0.86	23.21
North Central	1.69	1.99	2.20	3.54	4.33	2.89	2.02	2.09	3.69	3.05	2.12	1.78	31.39
Eastern	3.12	3.19	3.46	4.83	4.68	3.63	2.80	2.56	4.30	3.51	3.64	3.61	43.33
Trans-Pecos	0.42	0.44	0.34	0.42	1.02	1.46	2.18	2.05	2.16	1.13	0.52	0.39	12.53
Edwards Plateau	0.87	1.21	0.99	2.07	2.97	2.24	1.69	2.27	3.01	2.50	1.22	0.92	21.96
South Central	1.97	2.38	1.53	3.12	4.17	3.46	1.95	2.70	4.82	3.20	2.54	2.13	33.97
Upper Coast	3.20	3.06	2.29	3.36	4.45	4.37	3.83	4.25	6.06	3.48	3.58	3.61	45.54
Southern	1.00	1.24	0.71	1.78	2.92	2.63	1.49	2.50	4.34	2.56	1.27	1.00	23.44
Lower Valley	1.31	1.35	0.61	1.62	2.40	2.95	1.65	2.89	5.19	3.10	1.44	1.10	25.61

with a strong cold front moving through the state. Most of this precipitation is limited to the coastal, eastern and south central regions of the state where moisture is plentiful during the winter season.

Precipitation extremes at individual stations are of considerable significance. Extreme dryness, when it persists for an extended time, can lead to drought conditions. Heavy precipitation, on the other hand, is responsible for flooding and flash flooding. Table 5.6 gives the highest monthly and daily precipitation totals observed in Texas since the turn of the century. Excessive rainfall such as that shown in the table is often responsible for large economic losses as well as personal injuries and deaths. Urban areas are particularly vulnerable to flooding where development has taken place in flood plains.

There are basically two types of flooding in Texas. The first involves persistent and prolonged rainfall for periods ranging from several hours to a day or more. As illustrated in Table 5.6, this type flood can occur in many areas of the state but is most frequent in the upper coast and southern climatic regions. The second type flood results from intense thunderstorms during the spring and summer which produce heavy rainfall over limited areas for periods ranging from several minutes to a few hours. Rainfall of three inches or more in half an hour is not unusual. These events may produce dangerous flash floods which are most frequent along the Balcones Escarpment in the south central region, and in the Trans-Pecos, High Plains, Low Rolling Plains and Edwards Plateau.

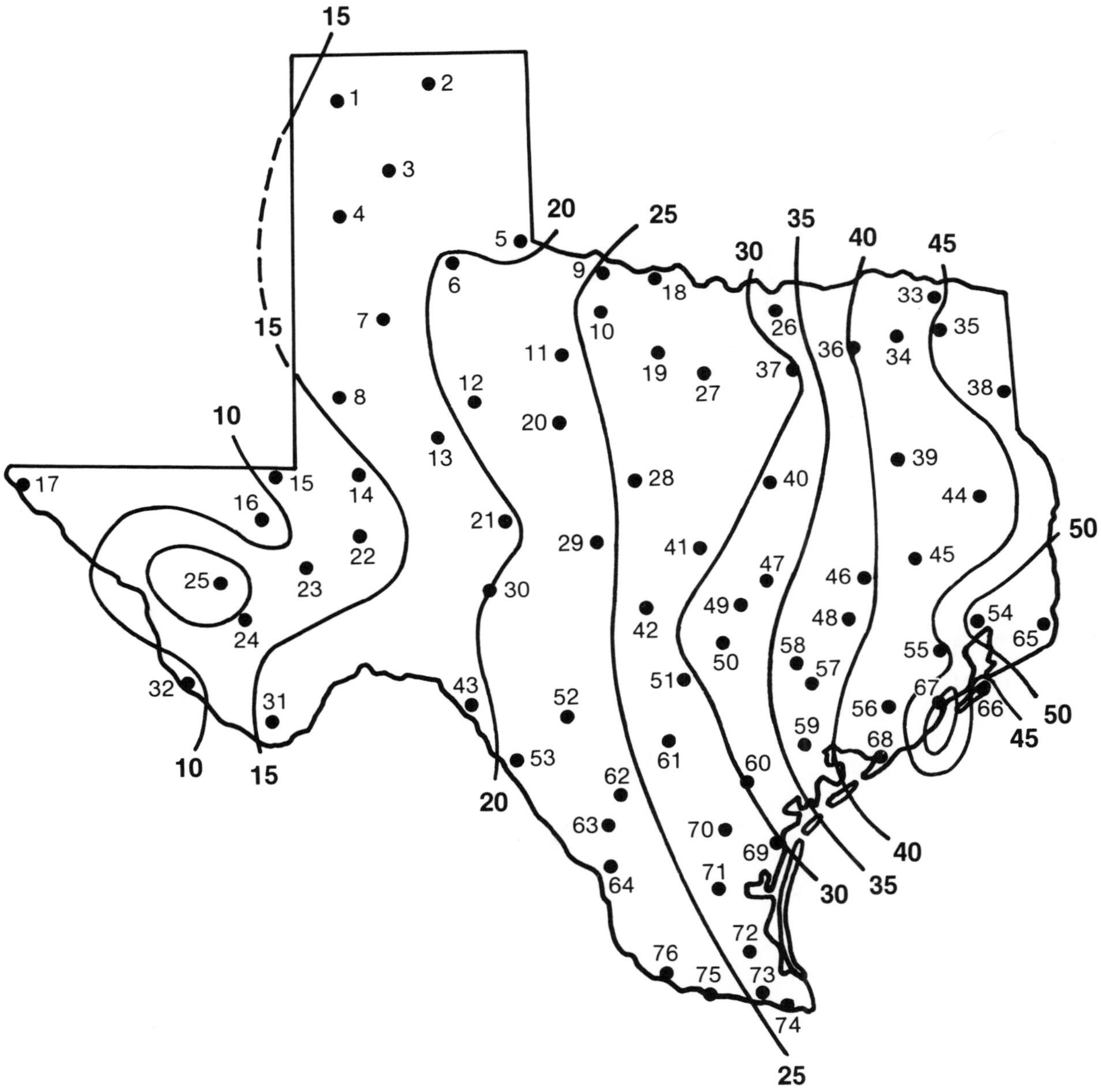

FIGURE 5.4
Average Annual
Precipitation

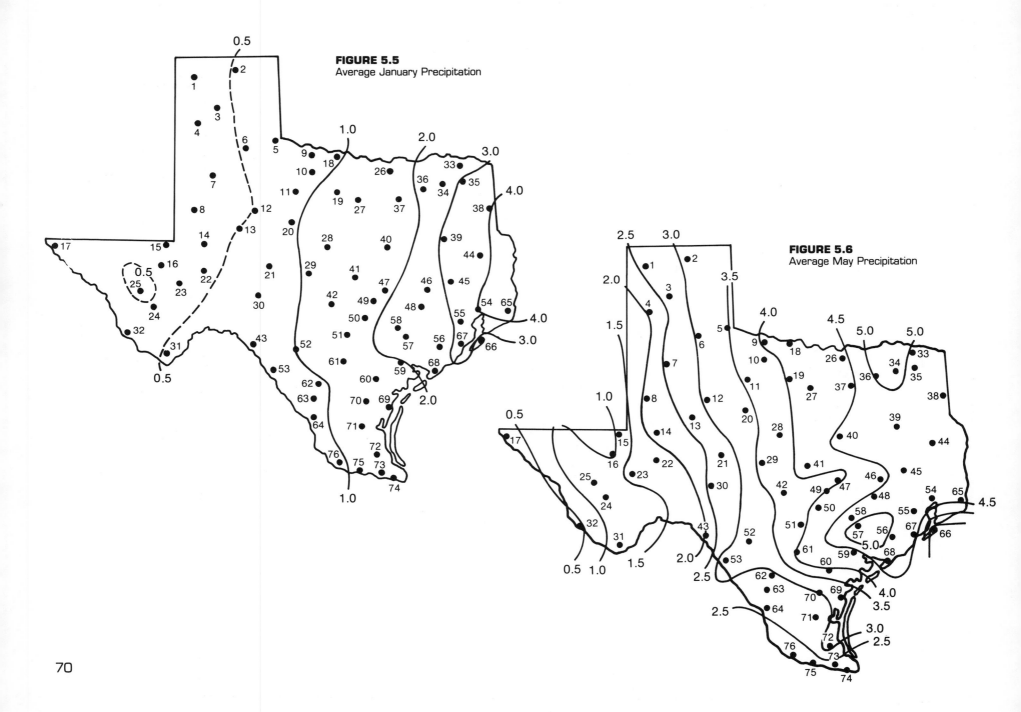

FIGURE 5.5
Average January Precipitation

FIGURE 5.6
Average May Precipitation

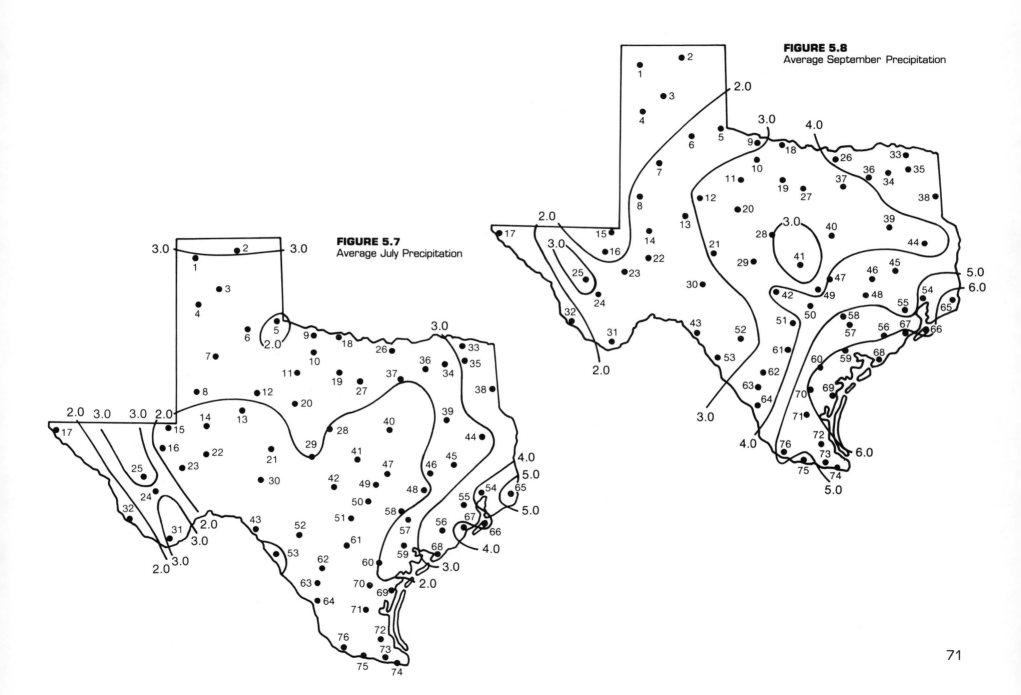

FIGURE 5.7
Average July Precipitation

FIGURE 5.8
Average September Precipitation

FIGURE 5.9
Average Monthly Precipitation
for Each of the Climatic
Regions

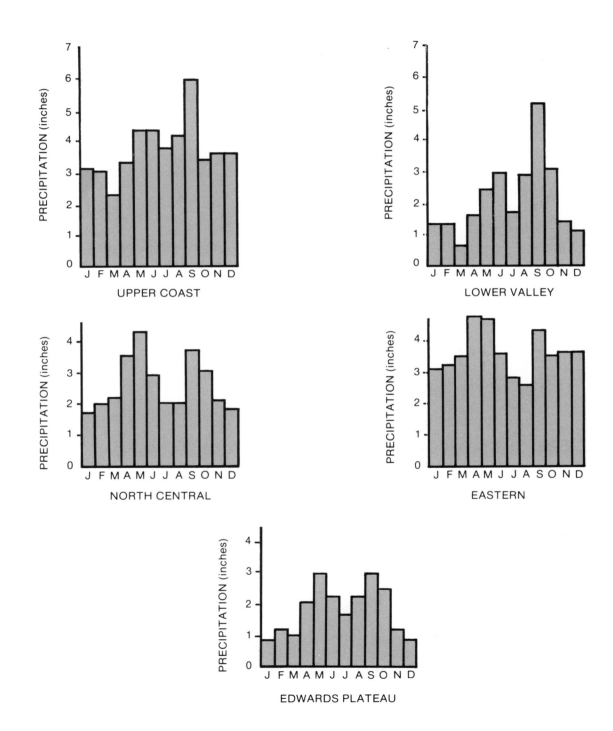

UPPER COAST

LOWER VALLEY

NORTH CENTRAL

EASTERN

EDWARDS PLATEAU

FIGURE 5.9
continued

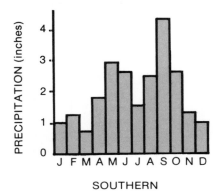

SOUTHERN

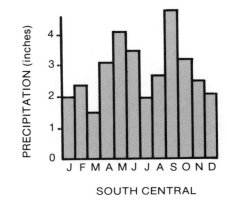

SOUTH CENTRAL

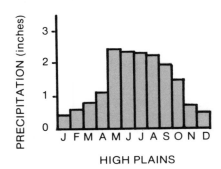

HIGH PLAINS

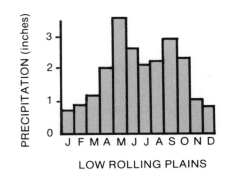

LOW ROLLING PLAINS

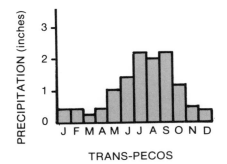

TRANS-PECOS

TABLE 5.6
Highest One-Month and
24-Hour Precipitation

Daily (24-hr) Precipitation

Amount (Inches)	Station	Climatic Region	Date		
29.05	Albany	North Central	4	Aug	1978
25.75	Alvin	Upper Coast	26	Jul	1979
21.02	Kaffie Ranch	Southern	12	Sep	1971
20.70	Hye	Edwards Plateau	11	Sep	1952
20.60	Montell	Edwards Plateau	27	Jun	1913
20.60	Deweyville	Upper Coast	18	Sep	1963
19.29	Danevang	Upper Coast	27-28	Aug	1945
19.20	Benavides No. 2	Southern	11	Sep	1971
19.03	Austin	South Central	9-10	Sep	1921
17.76	Port Arthur	Upper Coast	27-28	Jul	1943
17.47	Beaver	Edwards Plateau	11	Sep	1952
16.72	Freeport 2 NW	Upper Coast	26	Jul	1979
16.05	Smithville	South Central	20	Jun	1940
16.02	Hills Ranch	South Central	10	Sep	1921
16.02	Pandale	Low Rolling Plains	27	Jun	1954
16.00	Hemstead	Eastern	24	Nov	1940
15.87	Anahuac	Upper Coast	27-28	Aug	1945
15.00	Orange	Upper Coast	18	Sep	1963
15.71	Matagorda	Upper Coast	1	May	1911
15.69	Whitsett	Southern	22	Sep	1967
15.65	Houston	Upper Coast	27-28	Aug	1945
15.60	Eagle Pass	Southern	29	Jun	1936

Monthly Precipitation

Amount (Inches)	Station	Climatic Region	Date	
35.70	Alvin	Upper Coast	Jul	1979
32.78	Falfurrias	Southern	Sep	1967
31.61	Freeport	Upper Coast	Sep	1979
31.19	Albany	North Central	Aug	1978
30.95	Freeport	Upper Coast	Jul	1979
29.76	Port Lavaca	Upper Coast	Jun	1960
29.22	Aransas Pass	South Central	Sep	1967
29.19	Whitsett	Southern	Sep	1967
28.96	Deweyville	Upper Coast	Oct	1970
27.89	Kaffie Ranch	Edwards Plateau	Sep	1971
27.65	San Angelo	Edwards Plateau	Sep	1936
26.86	Port Arthur	Upper Coast	Jul	1979
26.79	San Augustine	Eastern	Aug	1915
26.68	Gladewater	Eastern	Apr	1966
26.31	Beaumont	Upper Coast	Oct	1970
26.30	Refugio	South Central	Sep	1971
26.06	Rio Grande City	Southern	Sep	1967
26.00	Cibolo Creek	South Central	Sep	1967
25.87	Taylor	North Central	Sep	1921
25.67	Pandale	Edwards Plateau	Jun	1954
25.59	Sinton	South Central	Sep	1967
25.57	Hemstead	Eastern	Nov	1940
25.54	New Gulf	Upper Coast	Jun	1960

DROUGHT

There are three types of regions in which water from precipitation is deficient. There are the great deserts and arid regions of the world in which a perpetual water shortage exists, regions in which the distribution of rainfall causes marked deficits of water to occur seasonally, and the usually well-watered areas which periodically experience water shortages due to lack of precipitation. An example of the first is the arid region of far West Texas where annual precipitation is normally below 10 inches. Although a good example of the second type cannot be found in Texas, a reasonable approximation is the High Plains where precipitation is limited to the warm season. It is the third type which is of particular interest—when expected precipitation does not occur. When this condition persists for an extended time, it is called a drought. The major drought of the 1930's and early 1950's which extended throughout much of the Great Plains of the United States had a profound effect on Texas. More limited droughts have periodically affected the state interfering with plant growth and water supplies and often leading to huge economic losses.

A definition of drought actually depends upon water demand in an area. From a meteorological standpoint, drought is simply a lack of precipitation. It occurs over limited areas of the globe for limited periods of time. As discussed in Chapter 3, areas associated with low pressure systems usually receive precipitation whenever sufficient moisture is available. High pressure areas, on the other hand, tend to suppress precipitation. When high pressure stagnates over a particular area for an extended time, drought results. These circumstances led to the Great Plains drought of the 30's and 50's as high pressure dominated the circulation over much of the country during times when precipitation was normally provided. Semiarid regions such as the High Plains and portions of the Edwards Plateau that are on the boundary between regions of deficit and excess rainfall are areas particularly sensitive to drought. When the dry period persists, all areas are susceptible.

Much of Texas is subject to recurring drought. It has occurred in the past and will in the future. Drought is an unusual event, but is not unexpected.

Appendix 1 / GLOSSARY

Absolute humidity	The number of grams of water vapor in a unit volume (cubic centimeter) of air.
Absolute temperature scale	(see Kelvin temperature scale)
Absolute zero	The temperature at which all molecular motion ceases.
Absorption	The process by which radiation is retained by a body or substance and transformed to heat.
Air mass	A large body of air which is horizontally homogeneous with respect to temperature and water vapor content.
Altocumulus	A cumulus-type cloud at middle levels which often exhibits a wavy or cellular appearance.
Altostratus	A stratus-type cloud at middle levels which has a sheetlike or layered appearance.
Anemometer	An instrument for measuring wind speed.
Aneroid barometer	An instrument for measuring atmospheric pressure which contains no liquid.
Anticyclone	An area of high pressure around which winds spiral clockwise in the northern hemisphere and counterclockwise in the southern hemisphere.
Barometer	An instrument for measuring atmospheric pressure.
Blue norther	A strong north wind.
Boiling point	The temperature at which a substance changes phase from liquid to gas.
Bouyant air	A parcel of air is bouyant if it is lighter than the surrounding air which gives it a tendency to rise.
Celsius temperature scale	A scale for measuring temperature on which the boiling and freezing points of water are 100° and 0°, respectively.
Centigrade temperature scale	(see Celsius temperature scale)
Cirrocumulus	A high level cumuliform cloud composed of ice.
Cirrostratus	A high level stratiform cloud composed of ice.
Cirrus	A high level cloud composed of ice.

Climate	The long-term summation of weather for a particular location or area.
Cloud	The visible result of condensation or sublimation in the free atmosphere.
Cold front	A front which moves such that cold air is replacing warmer air at the ground.
Condensation	The transition of a substance from vapor to liquid.
Conduction	The transfer of heat from molecule to molecule within a substance or from one substance to another.
Continental air	Air which is relatively dry because of a source region over a continent rather than the ocean.
Convection	The transfer of heat by movement of the heat-containing substance. Also used in meteorology to indicate the vertical movement of air.
Convergence	A flow of air into a particular region.
Coriolis effect	A deflection of the wind direction, to the right in the northern hemisphere and to the left in the southern hemisphere, due to the rotation of the earth.
Cumulonimbus	A thunderstorm cloud with extensive vertical development.
Cumulus	Well-defined individual clouds which grow in the vertical.
Cumulus stage	The initial stage in the development of a thunderstorm.
Cyclone	An area of low pressure around which winds spiral counterclockwise in the northern hemisphere and clockwise in the southern hemisphere.
Dew	Condensation on grass and other objects near the ground due to cooling.
Dew point	The temperature to which air must be cooled in order for condensation to occur.
Dissipating stage	The final stage in the life cycle of a thunderstorm.
Divergence	A flow of air out of a particular area.
Drizzle	Small drops between 0.2 and 0.5 mm in diameter that fall slowly and reduce visibility.
Dry line	A boundary which separates dry air from moist air and is often associated with the development of thunderstorms during the warm season.

Dust devil	A small, rapidly rotating whirlwind which develops on hot afternoons and is made visible by the dust, sand and/or debris it picks up.
Dust storm	High winds which cause sufficient blowing dust to significantly reduce visibility.
Evaporation	The transition of a substance from liquid to gas.
Fahrenheit temperature scale	A scale for measuring temperature on which the freezing and boiling points of water are 32° and 212° respectively.
Fog	A cloud in contact with the ground which reduces visibility to below one kilometer.
Freezing rain or drizzle	Rain or drizzle which freezes on contact with an object or the ground.
Front	A boundary separating two distinct air masses.
Frost	The deposition of frozen water resulting from cooling the air to the dewpoint when the dewpoint is below freezing.
Gravity	The force of attraction exerted by the earth.
Hail	Precipitation in the form of chunks of ice produced by thunderstorms.
Heat	A form of energy transferred from one substance to another as a result of a temperature difference.
High	An anticyclone.
Humidity	A general term that refers to the water vapor content of the air.
Hurricane	A severe tropical cyclone with winds exceeding 74 mph.
Hygrometer	An instrument for measuring the water vapor content of the air.
Isobar	A line connecting points of equal pressure.
Isohyet	A line connecting points with equal amounts of precipitation.
Isotherm	A line connecting points of equal temperature.
Jet stream	Strong winds concentrated within a narrow band.
Kelvin temperature scale	A temperature scale based upon the absolute zero of temperature, the point at which all molecular motion ceases.
Knot	A unit of speed equal to one nautical mile per hour.
Lapse rate	The rate of decrease of temperature with height.

Latent heat	Heat released or absorbed by a substance when it undergoes a change of phase (ex. evaporation, condensation).
Lightning	An electrical discharge characteristic of thunderstorms.
Low	A cyclone.
Maritime air	Air containing water vapor because of its source region over the ocean.
Mature stage	The middle stage of thunderstorm development characterized by both upward and downward motion.
Mesopause	The top of the mesosphere.
Mesosphere	The atmospheric region between the stratosphere and thermosphere (50-80 km) within which temperature decreases with height.
Meteorology	The science of the atmosphere.
Millibar	A unit of pressure equal to 1,000 dynes per square centimeter.
Nimbostratus	A stratus cloud from which precipitation is falling.
Orographic lifting	Lifting of air by mountains.
Polar air mass	A cold air mass having its source region in polar latitudes.
Polar easterlies	High latitude easterly winds.
Polar front	The non-continuous boundary separating polar and tropical air masses.
Precipitation	Any form of water which falls from the atmosphere and reaches the ground.
Pressure	The weight of the air over a particular surface area.
Prevailing westerlies	The region of westerly winds at middle latitudes.
Psychrometer	An instrument consisting of a wet- and dry-bulb thermometer used for measuring the water vapor content of the air.
Rain	Precipitation in the form of liquid water drops larger than drizzle.
Rain gauge	An instrument for measuring rainfall at a point.
Rain shadow	The region on the leeside of a mountain where precipitation is reduced.
Relative humidity	The ratio of the amount of water vapor in the air to the amount possible at a particular temperature and pressure.

Saturation	The condition expressed by a relative humidity of 100%.
Sea breeze	A wind blowing from ocean toward land.
Severe thunderstorm	A thunderstorm likely to produce hail, damaging winds and possibly tornadoes.
Sleet	Falling rain which freezes in the atmosphere before reaching the ground.
Snow	Precipitation in the form of ice crystals which forms by a transition of vapor to ice.
Squall line	A line of thunderstorms.
Stable	An atmosphere which tends to suppress upward vertical motion.
Storm surge	The tide produced by strong winds along a coastline and usually the result of tropical storms and hurricanes.
Stratopause	The top of the stratosphere.
Stratosphere	The atmospheric region above the troposphere [10-50 km] in which temperature increases with height.
Stratus	A low cloud layer usually with a uniform base.
Stationary front	A stationary boundary between warm and cold air.
Temperature	That property which determines whether one substance is in thermal equilibrium with another.
Thermal equilibrium	Two bodies in thermal equilibrium are characterized by no heat transfer between them.
Thermograph	An instrument that measures and records air temperature.
Thermometer	An instrument for measuring temperature.
Thermosphere	The atmospheric region above the mesosphere in which temperature increases with height.
Thunder	The sound produced as a result of the heating by lightning.
Thunderstorm	A localized storm accompanied by lightning and thunder.
Tornado	A violently rotating column of air in contact with the ground.
Trade winds	A system of low-level winds in the tropics.
Tropical air mass	A warm air mass whose source is in tropical regions.
Tropical cyclone	A cyclone originating over tropical oceans.

Tropopause	The top of the troposphere.
Troposphere	That region of the atmosphere next to the ground (0-10 km) in which temperature decreases with height.
Unstable atmosphere	An atmosphere which supports upward vertical motion.
Wall cloud	A lowering of the cloud base of a severe thunderstorm from which a funnel cloud or tornado may develop.
Warm front	A front which moves such that warm air is replacing cold air at the ground.
Water vapor	The gaseous form of water.
Wind	Air in motion relative to the earth's surface.

Appendix 2 / STATION LOCATIONS

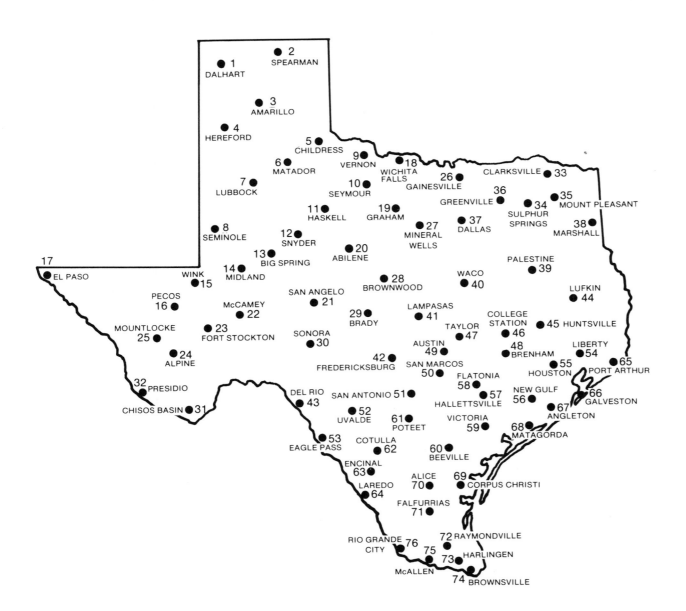

- 1 DALHART
- 2 SPEARMAN
- 3 AMARILLO
- 4 HEREFORD
- 5 CHILDRESS
- 6 MATADOR
- 7 LUBBOCK
- 8 SEMINOLE
- 9 VERNON
- 10 SEYMOUR
- 11 HASKELL
- 12 SNYDER
- 13 BIG SPRING
- 14 MIDLAND
- 15 WINK
- 16 PECOS
- 17 EL PASO
- 18 WICHITA FALLS
- 19 GRAHAM
- 20 ABILENE
- 21 SAN ANGELO
- 22 McCAMEY
- 23 FORT STOCKTON
- 24 ALPINE
- 25 MOUNTLOCKE
- 26 GAINESVILLE
- 27 MINERAL WELLS
- 28 BROWNWOOD
- 29 BRADY
- 30 SONORA
- 31 CHISOS BASIN
- 32 PRESIDIO
- 33 CLARKSVILLE
- 34 SULPHUR SPRINGS
- 35 MOUNT PLEASANT
- 36 GREENVILLE
- 37 DALLAS
- 38 MARSHALL
- 39 PALESTINE
- 40 WACO
- 41 LAMPASAS
- 42 FREDERICKSBURG
- 43 DEL RIO
- 44 LUFKIN
- 45 HUNTSVILLE
- 46 COLLEGE STATION
- 47 TAYLOR
- 48 BRENHAM
- 49 AUSTIN
- 50 SAN MARCOS
- 51 SAN ANTONIO
- 52 UVALDE
- 53 EAGLE PASS
- 54 LIBERTY
- 55 HOUSTON
- 56 NEW GULF
- 57 HALLETTSVILLE
- 58 FLATONIA
- 59 VICTORIA
- 60 BEEVILLE
- 61 POTEET
- 62 COTULLA
- 63 ENCINAL
- 64 LAREDO
- 65 PORT ARTHUR
- 66 GALVESTON
- 67 ANGLETON
- 68 MATAGORDA
- 69 CORPUS CHRISTI
- 70 ALICE
- 71 FALFURRIAS
- 72 RAYMONDVILLE
- 73 HARLINGEN
- 74 BROWNSVILLE
- 75 McALLEN
- 76 RIO GRANDE CITY

Appendix 3 / DIMENSIONS AND UNITS

Length [L]

1 kilometer [km] = 1000 meters [m]
3281 feet [ft]
0.62 mile [mi]

1 mile [mi] = 5280 feet [ft]
1609 meters [m]
1.61 kilometers [km]

1 meter [m] = 100 cm
3.28 ft
39.37 in

1 centimeter [cm] = 0.39 inch [in]
0.01 meter [m]

1 micrometer [μm] = 0.001 cm
0.000001 m

1 inch [in] = 2.54 cm
0.08 ft

Area [L X L] = L²

1 square centimeter [cm²] = 0.15 in²
1 square inch [in²] = 6.45 cm²
1 square meter [m²] = 10.76 ft²
1 square foot [ft²] = 0.09 m²

Volume [L X L X L] = L³

1 cubic centimeter [cm³] = 0.06 in³
1 cubic centimeter [in³] = 16.39 cm³
1 liter [ℓ] = 1000 cm³

Mass

1 gram [g] = 0.035 ounce
0.002 lb

1 kilogram [kg] = 1000 g
2.2 lb

Speed

1 knot [kt] = 1.15 miles per hour [mph]
0.51 meters per second [mps]
1.85 kilometers per hour [kph]

1 mph = 0.87 kt
0.45 mps
1.61 kph

1 kph = 0.54 kt
0.62 mph
0.28 mps

1 mps = 1.9 kt
2.2 mph
3.6 kph

Force

1 dyne = 2.2481 X 10⁻⁶ lb
1 lb = 4.448 X 10⁵ dyne
1 newton [N] = 10⁵ dyne

Energy

1 erg = 1 dyne cm = 2.388×10^{-8} calories (cal)

1 joule (J) = 10^7 erg = 0.239 cal

1 cal = 4.1855×10^7 erg = 4.1855 J

Pressure

1 millibar (mb) = 1000 dyne/cm²
                  0.75 millimeter of mercury
                  0.03 inch of mercury
                  0.01 pound per square inch
                  100 pascals (Pa)

1 lb/in² = 68.95 mb
              5.1715 centimeters of mercury
              2.036 inches of mercury
              68,947 dyne/cm

1 dyne/cm² = 1.4504×10^{-5} lb/in²
                    2.9530×10^{-5} inches of mercury
                    7.506×10^{-5} centimeters of mercury

1 bar = 10^5 N/m²
          10^6 dyne/cm²

1 inch of mercury = 0.491 lb/in²
                      33,864 dyne/cm²
                      33.864 mb

1 centimeter of mercury = 0.1934 lb/in²
                            13,332 dyne/cm²
                            13.332 mb

Appendix 4 / PLOTTING MODEL FOR UPPER AIR

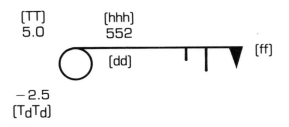

Symbols

		Example
dd:	Arrow shaft shows wind direction	90° or East
ff:	Barbs show wind speed (triangle, 50 knots; full barb, 10 knots; half barb, 5 knots)	65 Knots
TT:	Temperature (°C)	5.0°C
T_dT_d:	Dew point temperature (°C)	−2.5°C
hhh:	Height of pressure surface in meters with only last three digits given	1,552 m

Appendix 5 / TEMPERATURE AVERAGES (DEG F)

Station	Jan	Feb	Mar	Apr	May	Jun	Jul	Aug	Sep	Oct	Nov	Dec	Ann
Abilene	43.3	47.9	55.6	65.2	72.5	80.5	84.1	83.1	76.0	65.9	53.4	46.4	64.5
Alice	55.1	58.7	65.7	73.3	78.3	83.1	85.2	85.2	81.2	73.3	64.3	57.7	71.8
Alpine	46.3	48.9	54.9	62.9	70.0	76.7	76.8	75.5	70.7	62.9	52.7	47.4	62.1
Amarillo	35.4	39.6	46.4	56.5	65.5	74.9	78.8	77.0	69.7	59.2	45.4	38.3	57.2
Angleton	52.6	55.1	61.3	68.5	74.4	79.8	82.1	81.7	78.3	69.9	61.3	55.4	68.4
Austin	49.1	53.2	60.5	68.7	74.9	81.6	84.7	84.5	79.2	69.8	58.7	52.1	68.1
Beeville	53.0	56.2	63.4	70.9	76.2	81.4	83.9	83.9	79.6	71.5	62.1	55.7	69.8
Big Spring	43.4	47.5	55.2	64.8	72.5	80.2	82.8	81.7	74.9	65.4	52.9	46.3	64.0
Brady	44.4	48.2	55.6	65.1	71.7	79.0	82.7	81.8	75.6	65.7	53.8	47.0	64.2
Brenham	49.3	52.8	60.0	68.3	75.0	81.3	84.5	84.5	79.3	70.0	59.3	52.4	68.1
Brownsville	60.3	62.8	68.6	74.9	79.2	82.6	84.1	84.1	81.4	75.3	67.7	62.3	73.6
Brownwood	44.0	48.4	56.3	66.3	73.3	81.0	84.7	83.8	77.2	66.8	54.4	47.2	65.3
Childress	39.0	43.7	51.2	62.0	70.3	79.3	83.3	81.8	74.0	63.3	50.1	42.3	61.7
Chisos Basin	48.3	50.7	56.8	64.5	70.6	75.2	74.6	73.3	69.2	62.9	54.0	49.1	62.4
Clarksville	41.5	45.8	53.4	62.9	70.5	77.9	82.0	81.2	75.0	64.5	52.9	44.8	62.7
College Station	49.2	53.0	59.9	68.0	74.7	81.1	84.2	84.0	78.7	69.3	58.6	51.9	67.7
Corpus Christi	56.3	59.3	65.9	73.0	78.1	82.7	84.9	85.0	81.5	74.0	65.0	59.1	72.1
Cotulla	53.3	57.7	65.3	73.3	78.9	84.3	86.5	86.1	81.3	72.7	62.2	55.6	71.4
Dalhart	33.6	38.0	44.1	54.3	63.5	73.5	77.6	75.6	67.7	56.7	43.2	36.4	55.4
Dallas-Fort Worth	44.0	48.5	56.1	65.9	73.7	82.0	86.3	85.5	78.6	67.9	55.6	47.8	66.0
Del Rio	50.8	55.6	63.4	71.8	77.6	83.6	86.0	85.3	80.3	70.8	59.4	52.5	69.8
Eagle Pass	51.0	55.9	63.9	72.8	78.4	84.4	86.8	86.2	81.1	71.6	60.1	52.9	70.4
El Paso	44.2	48.4	55.1	63.6	71.9	80.8	82.5	80.3	74.1	63.6	51.4	44.4	63.4
Encinal	52.7	56.8	64.7	73.4	78.8	84.0	86.1	85.7	81.1	72.2	61.8	54.8	71.0
Falfurrias	55.8	59.2	66.5	74.3	79.0	83.5	85.5	85.0	81.1	73.3	63.9	57.9	72.1
Flatonia	51.5	55.1	62.2	69.6	75.2	80.9	83.9	83.9	79.0	70.9	60.7	54.3	68.9
Fort Stockton	45.5	49.0	55.6	64.9	72.5	79.9	81.5	80.2	74.4	65.0	53.6	47.3	64.1
Fredericksburg	48.1	51.7	59.1	66.9	72.7	79.0	81.8	81.3	76.2	67.2	56.4	50.5	65.9
Gainesville	40.3	45.0	52.9	63.1	70.9	79.4	84.0	83.2	75.8	64.8	52.2	44.1	63.0
Galveston	53.6	55.6	61.4	69.1	75.7	81.2	83.2	83.2	80.0	72.7	63.0	56.8	69.6
Graham	41.5	46.2	54.2	64.4	71.7	80.1	84.6	83.8	76.5	65.4	52.9	45.0	63.9

Station	Jan	Feb	Mar	Apr	May	Jun	Jul	Aug	Sep	Oct	Nov	Dec	Ann
Greenville	41.2	45.5	53.1	63.1	70.8	78.9	83.1	82.6	76.1	65.3	53.0	44.9	63.1
Hallettsville	52.6	56.0	63.1	70.1	75.8	81.4	84.1	84.2	79.6	71.1	61.4	55.1	69.5
Harlingen	58.7	61.6	68.1	75.0	79.2	82.7	84.2	84.6	81.2	74.6	66.5	61.0	73.1
Haskell	41.2	45.6	53.4	64.5	72.2	80.7	84.6	83.6	76.0	65.2	52.4	44.7	63.7
Hereford	35.5	39.0	45.5	55.5	64.3	73.6	77.0	75.1	68.1	57.8	44.9	38.1	56.2
Houston	51.4	54.5	61.0	68.7	74.9	80.6	83.1	82.6	78.4	69.7	60.1	54.0	68.3
Huntsville	48.4	52.0	59.3	67.7	74.4	80.6	83.6	83.1	77.8	68.9	58.3	51.4	67.1
Lampasas	44.6	48.4	55.7	65.0	71.7	79.4	83.2	82.5	76.3	65.9	54.2	47.5	64.5
Laredo	56.2	60.4	68.3	76.1	81.1	85.5	87.4	86.9	82.6	74.7	64.5	58.0	73.5
Liberty	50.7	53.7	60.5	68.2	74.8	80.3	82.9	82.6	78.5	69.2	59.6	53.2	67.9
Lubbock	38.8	42.6	50.2	60.3	69.0	77.6	79.8	77.9	71.2	61.0	48.5	41.5	59.9
Lufkin	48.6	52.0	59.3	67.6	74.3	80.3	83.3	82.9	78.0	68.2	57.4	51.0	66.9
Marshall	44.1	48.2	55.4	64.7	72.1	79.0	82.8	82.2	76.5	65.7	54.4	47.1	64.4
Matador	40.3	44.4	51.6	62.2	70.0	78.6	82.7	81.2	73.7	63.5	50.7	43.8	61.9
Matagorda	54.8	57.2	63.2	70.3	76.6	81.8	84.1	83.8	80.2	72.5	63.6	57.6	70.5
McAllen	58.8	61.8	68.6	75.4	79.5	83.0	84.4	85.0	81.9	74.9	66.5	60.7	73.4
McCamey	45.8	50.0	58.3	67.7	75.0	82.1	84.1	83.2	76.9	67.0	54.5	47.5	66.0
Midland	43.7	47.7	55.0	64.1	72.1	79.8	81.7	80.6	74.2	64.4	52.3	46.0	63.5
Mineral Wells	43.9	48.5	56.3	65.8	73.0	81.0	85.1	84.3	77.4	66.9	54.5	47.3	65.3
Mount Locke	42.7	44.5	50.3	58.2	65.0	71.2	70.6	69.2	65.2	58.7	49.2	43.9	57.4
Mount Pleasant	41.7	46.0	53.4	63.1	71.1	78.6	82.3	81.4	75.1	64.2	52.4	45.0	62.9
New Gulf	52.0	55.2	62.1	69.8	75.6	80.9	83.1	82.9	79.2	70.9	61.5	54.8	69.0
Palestine	45.5	49.4	56.9	65.8	72.8	79.5	83.0	82.7	77.0	67.5	56.1	48.8	65.4
Pecos	44.2	48.6	56.1	66.1	74.2	82.9	84.6	82.9	76.4	65.5	52.7	46.0	65.0
Port Arthur	51.9	54.9	61.4	69.0	75.6	81.2	83.1	82.8	79.2	70.2	60.6	54.7	68.7
Poteet	52.1	55.8	63.5	71.0	76.5	82.5	84.8	84.7	80.0	71.2	61.2	54.4	69.8
Presidio	50.4	55.0	62.9	71.9	79.7	87.2	87.4	85.6	80.4	70.5	58.0	50.6	70.0
Raymondville	58.3	61.2	68.0	75.3	79.3	82.9	84.6	84.9	81.5	74.5	66.3	60.5	73.1
Rio Grande City	56.5	60.2	68.1	76.4	80.7	84.9	86.6	86.7	82.3	73.9	64.6	58.3	73.3
San Angelo	45.5	49.7	57.5	66.8	73.9	81.4	84.3	83.3	76.5	66.6	54.6	48.0	65.7
San Antonio	50.4	54.3	61.8	69.6	75.5	81.9	84.6	84.2	79.4	70.2	59.5	53.0	68.7
San Marcos	48.4	52.3	59.4	67.7	74.0	80.3	83.2	83.2	78.1	68.8	57.9	51.0	67.0
Seminole	41.1	45.0	52.0	61.8	69.7	78.0	80.0	78.5	72.2	62.1	50.0	43.4	61.2
Seymour	39.6	44.1	51.9	63.2	71.3	80.1	84.5	83.6	75.5	64.2	50.9	43.0	62.7
Snyder	40.4	44.7	52.3	62.9	70.9	79.2	82.1	80.9	73.7	63.2	50.6	43.5	62.0
Sonora	46.8	51.0	59.0	67.3	73.8	79.8	82.1	81.2	75.8	66.2	55.1	48.5	65.6
Spearman	34.3	39.4	46.3	57.2	66.2	75.7	80.1	78.3	70.6	59.3	44.9	37.6	57.5

Station	Jan	Feb	Mar	Apr	May	Jun	Jul	Aug	Sep	Oct	Nov	Dec	Ann
Sulphur Springs	41.9	46.2	53.3	62.9	70.6	78.1	82.5	82.0	75.5	65.0	53.1	45.7	63.1
Taylor	46.9	50.8	58.5	67.4	74.0	80.9	84.4	84.3	78.5	68.6	57.1	50.0	66.8
Uvalde	50.7	55.0	62.7	70.7	76.3	82.1	84.2	83.6	79.3	70.4	59.7	52.8	69.0
Vernon	41.8	46.6	54.4	64.9	72.7	81.5	85.9	84.4	76.8	65.8	52.8	45.2	64.4
Victoria	53.4	56.5	63.3	70.9	76.7	82.0	84.5	84.2	80.1	71.9	62.3	56.1	70.1
Waco	46.2	50.5	58.1	67.1	74.2	81.9	85.9	85.6	79.2	68.8	57.0	49.5	67.0
Wichita Falls	40.3	45.3	53.3	63.7	71.8	80.7	85.6	84.3	76.2	65.1	52.0	43.9	63.5
Wink	43.5	48.1	55.7	65.4	73.7	81.7	83.4	82.1	75.6	64.9	52.1	45.1	64.3

Appendix 6 / PRECIPITATION AVERAGES (IN)

Station	Jan	Feb	Mar	Apr	May	Jun	Jul	Aug	Sep	Oct	Nov	Dec	Ann
Abilene	.97	.96	1.08	2.35	3.25	2.52	2.11	2.47	3.06	2.32	1.32	.85	23.26
Alice	1.31	1.52	.75	1.66	2.95	3.33	1.96	2.81	6.35	3.14	1.59	1.14	28.51
Alpine	.47	.43	.35	.35	1.03	1.95	2.91	2.58	2.61	1.16	.57	.42	14.83
Amarillo	.46	.57	.87	1.08	2.79	3.50	2.70	2.95	1.72	1.39	.58	.49	19.10
Angleton	3.79	3.72	2.90	3.15	4.63	5.49	4.54	4.99	6.97	3.51	4.25	4.34	52.28
Austin	1.60	2.49	1.68	3.11	4.19	3.06	1.89	2.24	3.60	3.38	2.20	2.06	31.50
Beeville	1.78	2.11	1.05	2.33	3.61	3.25	1.94	2.95	5.47	2.97	1.95	1.70	31.11
Big Spring	.58	.64	.75	1.42	2.77	1.88	1.85	1.88	2.87	1.74	.73	.61	17.72
Brady	1.06	1.36	1.19	2.30	3.60	2.17	2.07	2.35	3.67	2.48	1.35	1.06	24.66
Brenham	2.71	2.99	2.09	4.02	4.57	3.63	1.84	2.60	4.72	3.52	3.81	3.22	39.72
Brownsville	1.25	1.55	.50	1.57	2.15	2.70	1.51	2.83	5.24	3.54	1.44	1.16	25.44
Brownwood	1.38	1.44	1.52	2.69	3.88	2.89	1.66	1.95	3.00	2.96	1.50	1.23	26.10
Childress	.57	.71	1.07	1.76	3.50	2.60	1.95	1.91	2.32	2.04	.76	.70	19.89
Chisos Basin	.54	.46	.37	.53	1.47	1.85	3.16	3.04	2.74	1.45	.58	.52	16.71
Clarksville	2.67	3.13	4.01	5.22	4.96	3.52	3.03	2.25	4.07	3.93	3.91	3.41	44.11
College Station	2.48	2.97	2.39	4.34	4.35	3.21	2.39	2.30	4.93	3.42	3.33	2.97	39.08
Corpus Christi	1.58	1.68	.90	1.93	3.23	3.13	1.54	3.24	6.06	3.33	1.61	1.45	29.68
Cotulla	.78	1.14	.76	1.93	2.92	2.22	1.13	2.66	3.19	2.80	1.13	.99	21.65
Dalhart	.38	.45	.73	1.18	2.61	2.11	2.98	2.60	1.33	1.12	.58	.38	16.45
Dallas-Fort Worth	1.65	1.93	2.42	3.63	4.27	2.59	2.00	1.76	3.31	2.47	1.76	1.67	29.45
Del Rio	.51	.89	.63	1.85	1.99	1.72	1.69	1.60	2.73	2.24	.80	.55	17.19
Eagle Pass	.62	.81	.67	1.83	3.24	2.37	2.02	2.42	2.92	2.47	.88	.70	20.95
El Paso	.38	.45	.32	.19	.24	.56	1.60	1.21	1.42	.73	.33	.39	7.82
Encinal	.96	1.18	.77	1.69	2.57	2.34	1.22	2.31	3.78	2.44	1.20	.95	21.41
Falfurrias	1.40	1.35	.66	1.42	2.92	3.09	1.69	2.84	5.60	2.41	1.26	1.13	25.77
Flatonia	2.28	2.76	1.67	4.00	4.60	4.15	1.81	2.39	5.19	3.16	2.78	2.62	37.41
Fort Stockton	.42	.60	.43	.52	1.61	1.43	1.30	1.46	2.14	1.09	.77	.44	12.21
Fredericksburg	1.16	1.66	1.35	2.66	3.76	3.23	1.78	2.89	4.02	3.18	1.72	1.26	28.67
Gainesville	1.65	1.95	2.74	3.40	4.34	3.24	2.15	2.23	4.25	3.18	2.19	1.67	32.99
Galveston	2.96	2.34	2.10	2.62	3.30	3.48	3.77	4.40	5.82	2.60	3.23	3.62	40.24

Station	Jan	Feb	Mar	Apr	May	Jun	Jul	Aug	Sep	Oct	Nov	Dec	Ann
Graham	1.30	1.36	1.47	3.09	4.11	2.84	2.04	2.33	3.94	2.59	1.71	1.23	28.01
Greenville	2.28	2.67	3.40	4.73	5.30	3.24	2.73	2.14	4.47	3.68	3.04	2.75	40.43
Halletsville	2.38	2.63	1.97	3.38	5.22	3.83	2.39	2.83	5.29	3.07	2.91	2.50	38.40
Harlingen	1.38	1.46	.68	1.77	2.27	2.79	1.83	3.26	5.25	2.98	1.61	1.20	26.48
Haskell	.81	1.03	1.20	2.03	3.75	2.33	2.33	2.77	3.30	2.46	1.23	.89	24.13
Hereford	.38	.52	.70	.98	1.85	2.92	2.15	2.37	1.58	1.52	.64	.40	16.01
Houston	3.21	3.25	2.68	4.24	4.69	4.06	3.33	3.66	4.93	3.67	3.38	3.66	44.77
Huntsville	3.20	3.36	2.83	4.60	4.73	3.86	2.69	3.06	4.96	3.42	3.50	3.99	44.20
Lampasas	1.52	2.02	1.78	3.09	4.22	2.57	1.83	2.78	2.85	3.06	1.98	1.82	29.52
Laredo	.73	1.05	.40	1.22	2.54	2.74	1.06	2.67	3.45	2.18	1.23	.87	20.14
Liberty	4.00	3.93	2.66	4.35	4.31	4.46	4.19	3.84	5.34	4.22	4.59	4.76	50.65
Lubbock	.38	.57	.90	1.08	2.59	2.81	2.34	2.20	2.06	1.81	.59	.43	17.76
Lufkin	3.55	3.05	3.38	4.27	4.31	3.39	2.81	2.46	3.72	2.98	3.59	3.97	41.48
Marshall	4.14	3.72	3.99	5.10	4.86	3.73	3.31	2.42	4.11	3.18	3.74	4.11	46.41
Matador	.51	.70	.92	1.61	3.12	3.13	2.24	2.30	2.39	2.02	.79	.70	20.43
Matagorda	2.92	2.54	1.70	2.91	4.79	4.03	3.23	4.05	7.30	3.37	3.50	2.86	43.20
McAllen	1.25	1.10	.64	1.56	2.08	2.79	1.59	2.38	4.29	3.20	1.15	1.01	23.04
McCamey	.38	.55	.45	.71	1.75	1.34	1.28	1.46	2.25	1.48	.65	.43	12.73
Midland	.42	.58	.51	.84	2.05	1.44	1.72	1.60	2.08	1.41	.60	.45	13.70
Mineral Wells	1.61	1.68	1.99	3.41	4.06	2.59	2.27	2.18	3.30	2.93	1.82	1.43	29.27
Mount Locke	.60	.52	.43	.35	1.36	2.12	3.90	4.05	3.02	1.46	.65	.48	18.94
Mount Pleasant	3.11	3.38	3.88	5.20	4.52	3.71	3.27	2.71	4.35	3.62	4.11	3.64	45.50
New Gulf	2.68	2.74	2.01	2.94	4.95	4.93	3.65	4.32	5.77	3.53	3.13	2.97	43.62
Palestine	3.13	2.98	3.62	4.52	4.74	3.75	2.07	2.76	3.76	3.56	3.45	3.38	41.72
Pecos	.33	.39	.28	.43	.95	1.02	1.34	1.19	1.79	1.08	.46	.31	9.57
Port Arthur	4.18	3.71	2.93	4.05	4.50	3.96	5.37	5.45	6.13	3.63	4.33	4.55	52.79
Poteet	1.30	1.93	1.06	2.94	4.05	2.57	1.49	2.42	4.00	2.75	1.88	1.42	27.81
Presidio	.26	.31	.18	.26	.49	1.35	1.62	1.62	1.65	.78	.40	.28	9.20
Raymondville	1.36	1.29	.63	1.60	3.11	3.50	1.66	3.10	5.97	2.68	1.55	1.03	27.48
Rio Grande City	.81	.90	.57	1.50	2.28	2.02	1.30	2.16	5.11	2.28	.95	.69	20.57
San Angelo	.64	.84	.79	1.75	2.52	1.88	1.22	1.85	3.04	2.05	.97	.64	18.19
San Antonio	1.55	1.86	1.33	2.73	3.67	3.03	1.92	2.69	3.75	2.88	2.34	1.38	29.13
San Marcos	1.86	2.69	1.60	3.37	4.42	3.38	1.81	2.39	4.39	3.46	2.81	2.13	34.31
Seminole	.47	.61	.66	1.02	2.40	1.81	2.14	2.18	2.28	1.58	.64	.37	15.80
Seymour	.91	1.05	1.40	2.11	3.76	2.81	2.54	2.34	3.61	2.77	1.29	1.07	25.66
Snyder	.50	.64	.78	1.67	3.05	2.45	2.03	2.40	3.05	2.22	.91	.59	20.29

Station	Jan	Feb	Mar	Apr	May	Jun	Jul	Aug	Sep	Oct	Nov	Dec	Ann
Sonora	.72	1.13	.75	2.04	2.67	2.23	1.92	2.26	2.82	2.44	1.07	.65	20.70
Spearman	.52	.64	1.18	1.12	3.23	2.59	3.01	2.50	1.71	1.23	.92	.50	19.15
Sulphur Springs	2.65	2.93	3.57	5.37	5.01	3.89	2.80	2.55	4.53	3.97	3.47	3.42	44.16
Taylor	1.97	2.83	2.06	3.56	3.95	3.17	1.54	2.14	4.41	3.54	2.66	2.38	34.21
Uvalde	1.00	1.26	.95	2.17	3.26	2.81	1.74	2.94	2.73	3.08	1.19	.97	24.10
Vernon	.85	1.04	1.50	2.37	4.63	2.86	2.14	2.00	2.91	2.81	1.26	.88	25.25
Victoria	1.87	2.24	1.34	2.61	4.47	4.53	2.58	3.33	6.24	3.31	2.24	2.14	36.90
Waco	1.69	2.04	1.99	3.79	4.73	2.58	1.78	1.95	3.18	3.06	2.24	1.92	30.95
Wichita Falls	.93	1.00	1.82	2.99	4.34	2.85	2.00	2.14	3.41	2.61	1.42	1.22	26.73
Wink	.36	.36	.35	.74	1.05	1.43	1.60	1.24	1.90	1.27	.43	.29	11.02

Index